ELECTRONIC AND VIBRATIONAL PROPERTIES
OF POINT DEFECTS IN IONIC CRYSTALS

SERIES DEFECTS IN CRYSTALLINE SOLIDS

Editors:

S. AMELINCKX
R. GEVERS
J. NIHOUL

Studiecentrum voor Kernenergie, Mol,
and
University of Antwerp, Belgium

NORTH-HOLLAND PUBLISHING COMPANY
AMSTERDAM · NEW YORK · OXFORD

ELECTRONIC AND VIBRATIONAL PROPERTIES OF POINT DEFECTS IN IONIC CRYSTALS

Yves FARGE
Université de Paris – Sud

Marco P. FONTANA
Università degli Studi di Parma

1979

NORTH-HOLLAND PUBLISHING COMPANY
AMSTERDAM · NEW YORK · OXFORD

ISBN 0 444 85272 7

Publishers:

North-Holland Publishing Company — Amsterdam · New York · Oxford

Sole distributors for the USA and Canada:

Elsevier North-Holland, Inc.
52 Vanderbilt Avenue
New York, N.Y. 10017

Library of Congress Cataloging in Publication Data

Farge, Y
    Electronic and vibrational properties of point defects in ionic crystals.

    First published in 1974 under title: Perturbations électroniques et vibrationnelles localisées dans les solides ioniques.
    Includes bibliographies.
    1. Crystals-Defects.   2. Color centers.   3. Ionic crystals.   I. Fontana.
Marco P., joint author. II. Title.
QD921.F33 1979   548'.842   79–204
ISBN 0–444–85272–7

Printed in The Netherlands

PREFACE

Ionic materials make up a particular class of solids, characterized both by a marked localization of electrons and strong insulating properties. A variety of thermal and chemical treatments, or irradiations, easily create in the crystalline lattice of these solids so-called "point defects", i.e. impurity atoms, ion vacancies, interstitials. Electrons may in turn be trapped by such defects in strongly bound states, although the localization may not be as strong as for the intrinsic electronic states. The extra electron states are called "color centers" in these materials, which are in general, transparent in the visible spectrum, when no defects are present.

The experimental study of these color centers in alkali-halides, carried out mainly by German physicists, had brought a basic understanding of the main creation mechanisms of these defects and of their fundamental properties. This understanding became more precise and quantitative in the years following World War II, and about ten years ago one could think the subject mature. In fact, remarkable experimental and theoretical developments have completely renewed this subject just during the last fifteen years. These developments refer mainly to a refinement of spectroscopic techniques, accompanied by renewed interest in the optical properties of atoms and isolated molecules. A conspicuous research effort has brought better understanding of excited electronic states, particularly at ionization energies; furthermore a systematic exploration of electron–lattice coupling effects has been performed: vibronic structures, Jahn–Teller effects, localized and resonant vibrations, para-elasticity, etc ... have been investigated. These studies have also, given precious information about intrinsic electronic states, (band structures, excitons), and of vibrational spectra; such knowledge in turn was extended to the study of materials more complex than the alkali-halides.

This monograph mainly analyses these recent developments, after a brief introductory summary of some fundamental notions. Dr. Farge and Prof. Fontana, are particularly suited to the task, having themselves significantly contributed to these developments. This book is also based on a graduate course given in the Laboratoire de Physique des Solides, at Orsay. I think that it will be useful for the many physicists who, in industry or university laboratories, study or use these subtle but nonetheless interesting phenomena. I do hope also that whatever physicists have been able to learn in the "pure" case of alkali-halides will be useful in related fields, such as the luminescence of ionic-covalent compounds (sulphides, oxides), the physics of photochromic materials, or even the photographic process in silver salts!

J. Friedel

CONTENTS

INTRODUCTION

Most of the interesting properties of matter in the solid state are connected with the presence of defects and impurities which differentiate the real solid from the "perfect" solid. The perfect solid is a Platonic "idea" whose existence in science may indicate a certain philosophical attitude towards matter. More concretely, the perfect solid is just a model at the basis of a system which allows the testing of various theoretical approaches to the study of solids; the model helps in classifying in a controlled and quantitative manner the great variety of deviations from it shown by the real solid. Following the same line of thought, the study of simple solid, in which imperfections may be precisely controlled, is also very important, both from a fundamental and from an applied point of view.

It is evident that the definition of the perfect solid and that of a defect are inextricably related: a circular reasoning cannot be avoided, since a perfect solid is a solid without defects, and a defect is that which makes a solid imperfect.

The very first X-ray diffraction studies indicated that the majority of solids were composed of atoms arranged in more or less symmetric lattices. A perfect solid can then be defined as the solid composed of such lattices and in which the local symmetry around each and every atom is not perturbed in any way. Perturbations in an actual solid may be due to chemical impurities, interstitials, vacancies; furthermore, other perturbations may be due to internal surfaces, glide planes, dislocations, etc ... This last type of perturbation, which we shall call mechanical, is very important in determining many of the macroscopic properties of solids; however its influence on fundamental interactions of atoms on the microscopic level is marginal and therefore we shall not consider them in this book.

Generally speaking during the first half of this century, physics has passed from a study of the structure to a study of the dynamics of systems; for solids, we have changed our view of a crystal from that of a regular and static distribution of atoms to that of a gas of a variety of elementary excitations, in more or less strong interaction with each other. Thus, although an increasing number of properties of solids have been attributed to well defined defects, the theoretical study has shifted and the defect has been often used as a probe of the interactions between fundamental excitations in the solid state. In fact in many cases the introduction of a defect will give rise to new and interesting effects.

In this book we shall limit ourselves to defects in ionic solids. Historically these solids have been the first to be studied quantitatively and in fact some of the early triumphs of quantum mechanics were connected with the experimental and theoretical study of ionic solids. These solids are also interesting because they have simple crystal structures and in many cases it is easy to grow them with very low impurity content. Among the early works on the subject, we may cite Bragg's X-ray diffraction work and Born's book "Atomtheorie der Festen Zustandes". However defects in ionic solids were not studied quantitatively until the twenties; most of this early work was performed by Pohl and collaborators at Göttingen and the results are very well known. Pohl was intrigued by the fact that some samples of alkalihalides were colored, whereas they should have been transparent. The colored materials also had remarkable differences in some of their characteristics as compared with the non-colored materials. Furthermore, some of the colorations found in the natural crystals could be reproduced in the transparent crystals by irradiating them with X-rays.

The fundamental result of the work of Pohl is that the coloration is due to the presence of specific point defects in the crystal, called by Pohl "color centers" (Farbenzentren). The most important and well known of these centers became then known as the F-center, and most studies on defects in ionic solids have been concerned with this particular center.

Pohl and other researchers after him, have not only demonstrated that defects could cause substantial changes in the properties of the "pure" crystals: they also showed that it is possible to introduce defects in a controlled manner so that quantitative studies could be performed. Simultaneously with the study of color centers in alkali-halides, chemical impurity centers were also investigated. Also in this case, the simple structures and the high purity levels possible for alkali-halides (and similar ionic substances) make these materials ideal candidates for the quantitative study of luminescence or modification of electronic structure caused by impurities.

In the thirties, most work was performed on the F-center and the Tl-impurity center. Among the most important results, we may cite De Boer's model for the F-center (an electron trapped in an anionic vacancy), Seitz's model for optical transitions of the Tl center, the identification of several other color centers, and the beginning of an understanding of the production mechanisms of color centers by ionizing radiations. One of the most important consequences of these investigations was the understanding by solid state physicists of the considerable importance of defects in the behavior of solids.

This understanding was later used extensively in the study of semiconductors that led to the development of the transistor.

The study of defects in ionic solids was subsequently continued for two main reasons. Firstly, such a study could yield probe systems for theories and methods to be applied to more esoteric materials. Secondly, the development of atomic energy (and in particular of the atomic bomb) has motivated a certain number of civilian and military organizations to finance research in order to understand more quantitatively the effect of various types of radiation on matter. This situation has led to a certain isolation of this research domain; furthermore the researchers were led to study less well known substances for which quantitative results were very difficult to obtain. The work carried out in the fifties and early sixties is well described in the book by Schulman and Compton "Color Center of Solids". This book marks a turning point in the physics of color centers. During the fifties, the electron paramagnetic resonance technique was applied to color centers; also in this period, most of the important color centers were identified. Other important developments during this period were the detection of F-center luminescence, more quantitative studies on defect creation mechanisms, and a quantitative theory of optical transitions induced by defects.

Two factors have contributed to a radical change of emphasis in the research performed in the sixties and seventies; one is of course the detailed understanding of the structure of the most important defects; the other is related to the use of the very sensitive modulation techniques made possible by the development of the lock-in amplifier. By measuring the weak Faraday rotation or magnetic circular dichroism of the F-center by such modulation techniques, it was possible for instance, to obtain the value of the spin–orbit splitting in the excited state of the F-center. Other important results are the measurement of the radiative lifetime of the F-center and the relative theoretical model, the detection of Stark and Zeeman effects, both in absorption and emission for the F-center and other color and impurity centers; the detailed study of the F_A-center, the phenomena of para-elasticity and para-electricity due to the tunnelling states of some centers; the use of infrared and Raman spectroscopy, and of thermal conductivity measurements to study the perturbations in the vibrational dynamics of the crystal caused by the defects; the discovery of the optical analog of the Mössbauer effect, namely the so-called zero-phonon lines in the absorption and emission spectra of some color centers; the understanding of the complex band shapes of Tl (and similar) centers and of the complex phenomenology of phosphor emission.

This book will be dedicated essentially to the work on color center physics of recent years. Previous work is well covered in the books of Schulman and Compton, Markham, in the review articles of Seitz, Hilsch and Pohl. Some of the subject matter of this book is treated in detailed articles in the book "The Physics of Color Centers", (W. B. Fowler, ed.). We strongly encourage the interested reader to peruse these basic works (given in a short bibliography following this introduction) for a more detailed treatment of the various subjects. In the following pages we try to give a general introduction to the subject of defects in ionic solids for graduate students or colleagues from other domains of physics. As already stated, we shall discuss essentially the work of the last ten years and we shall expand on the results either of a general character or which feature a particularly interesting physical idea; we shall also try to give an idea of possible future developments of the subject and of possible connections with other research domains. Such a program has obliged us to discriminate in our choice of subjects. Our choices evidently reflect our preferences and opinions and we plead guilty in advance to accusations of partiality. We also would like to emphasize that this is not a review book on an already mature subject. We felt instead in writing it, that some of the most exciting results have given this domain a new lease of life and new research directions and it is towards these new frontiers that we would like to lead our reader.

We wish to thank G. Calas, W. B. Fowler, D. B. Fitchen, J. Friedel, P. Lagarde, F. Lüty and G. Toulouse, for advice and suggestions. Our thanks also goes to the several authors who have given authorization for us to reproduce figures from their published work, and to Mme Touchant and Mr. Saint-Martin, who have redesigned such figures.

General books on color centers in ionic solids

N. F. Mott and R. W. Gurney, Electronic Process in Ionic Crystals, 2nd edition (Oxford University Press, London, 1949).

F. Seitz, Rev. Mod. Phys. **18** (1946) 384; **26** (1954) 7.

J. H. Schulman and W. D. Compton, Color Centers in Solids (Pergamon Press, Oxford, 1962).

J. J. Markham, F-Centers in Alkali-Halides, Solid State Physics (1966) Suppl. 8.

W. B. Fowler (ed.) Physics of Color Centers (Academic Press, New York, 1968).

J. H. Crawford Jr. and L. F. Slifkin (ed.) Point Defects in Solids (Plenum Press, New York, Vol. 1 1972, Vol. 2 1975).

J. C. Kelly and P. D. Townsend, Colour Centers and Imperfection in Solids (Chatto and Windus, London, 1973).

C. P. Flynn, Point Defects and Diffusion (Oxford University Press, London, 1972).

R. C. Newman, Infrared Studies of Crystal Defects (Taylor and Francis, London, 1973).

B. Henderson, Defects in Crystalline Solids (Arnold, London, 1972).

W. Hayes, Crystals with Fluorite Structure (Clarendon Press, Oxford, 1974).

A. M. Stoneham, Theory of Defects in Solids (Clarendon Press, Oxford, 1975).

A. E. Hughes and D. Pooley, Real Solids and Radiation (Wykeham Publications, Taylor and Francis, London, 1975).

B. Henderson and A. E. Hughes (eds.) Defects and Their Structure in Non-Metallic Solids (Plenum Press, New York, (1976).

R. K. Watts, Point Defects in Crystals (Wiley Interscience, Chichester, 1977).

INTRODUCTION TO THE ENGLISH EDITION

This book was published in French in 1974 and evidently a lot of new results have appeared since that time. We have tried to introduce some of them, when they appeared to us to be fundamental enough and constituted a general basis for more complex systems. For example, picosecond laser measurements for the creation of F-centers, the Jahn–Teller effect of a triply degenerate orbital state (Tl^+ impurity in alkali halides), non-radiative transition processes using the triplet state of the F_2–centers, or the use of color centers to make tunable lasers. However our purpose was not to extensively cover a very broad field but to show that the field of color centers was still useful in generating some fundamental concepts which can be used in more complex systems. In this sense we have decided to keep the spirit of the original edition unaltered, although the English edition is expected to reach a wider audience. However some major changes were made: the lengthy didactic chapter 1 on the general properties of ionic crystals was suppressed and some of its more advanced subjects have been discussed in chapter 4 in an extended version i.e. the self-trapped exciton). The order of the chapters discussing electron–vibration interaction and electronic states have been reversed. Since what, at the time of writing the French edition, were avant-garde applications are now fairly standard techniques we have removed from the last chapter the paragraph on optical detection of magnetic resonances which found its proper place in chapter 4. Instead, in the applications chapter we placed the extended and up-dated discussion of radiation damage and defect creation mechanisms. These subjects have made great progress in recent years and constitute one of the best examples of the application of color center techniques and results.

Is the field of color centers closed as is sometimes said? We think that in fact, many questions are still unsolved. Amongst them, we could cite:

photochemical reactions in pure and doped crystals. As we have said previously our knowledge has greatly improved but we do not yet understand why and how an impurity or a normal ion can be ejected after an optical excitation (dynamics of excited electronic states).

Thermal aggregation and thermal bleaching of defects: our experimental techniques are very poor at determining and measuring large defects, large from the point of view of point defects, but still very small for electron microscopy or X-ray diffuse scattering.

Resonant Raman scattering and hot luminescence: only a few systems have been studied and other color centers could be used to test theoretical predictions.

However, it is clear that this field is mature enough for its results to be extrapolated to systems more complex than alkali halides or alkaline earth-oxides or fluorides. Color centers can be used as a local probe to study phase transitions (the F-center in KCN for example) in crystallographic or magnetic systems. They can also give important information on natural materials and their history etc ... Thus it is certainly important to continue their study in more complex materials. It is interesting in this respect to analyze the scientific activity of our colleagues who gave important contributions in the sixties and at the beginning of the seventies. Few of them remain directly connected with color centers and they are looking for important and unsolved questions. Some of them are now studying phase transitions using the ESR or optical techniques they elaborated for color centers. Some others are working on complex systems, solutions, biophysical systems using Raman scattering or other techniques. Many of them have jumped into synchrotron radiation which provides new methods to study the properties of insulators as well as metals or alloys; the interpretation of their results very often needs a localized description of excitations, a description which is familiar to them.

Writing this book, we just wanted to present as clearly as possible some fundamental concepts typical of the study of point defects in ionic crystals. At the same time we wanted to give as complete a picture as possible (given the scope and the size of the book) of the most interesting results in the field and accompanying theoretical models. Thus we hope that this book and the extensive bibliography it contains will serve as a general – but not superficial – introduction to the subject of localized states in insulators, intended especially for the advanced student who would be interested in doing research in this field or for the physicist or the physico-chemist who would want to apply the techniques and concepts characteristic of this field to his own research needs. In this sense we feel, and hope, that this book finds a useful place among other books on the same subject.

Prof. G. Viliani of the University of Trento (Italy) has contributed to this revised edition by collaborating in substantially re-writing chapters 3 and 4, where his knowledge of the Jahn–Teller problem was greatly exploited. We wish to express our gratitude to him.

1 COLOR CENTERS IN IONIC CRYSTALS: AN INTRODUCTION

Ionic crystals form a large class of materials. Alkali-halides are the simplest ones with alkaline earth oxides which crystallize in the well known NaCl structure. Alkaline earth halides are more complex, some of them have a simple structure like CaF_2, some others have a low symmetry like MgF_2; mixed crystals like BaClF can be considered from the point of view of barium–barium interactions like a two-dimensional material [1]. The class of magnetic ionic materials is also well known and can give magnetic systems having three dimensions (MnF_2, MnO . . .), two dimensions like K_2NiF_4 or one dimension like $CsNiF_3$ [2]. It is also well known that a majority of ferroelectric and antiferroelectric systems are ionic materials, the prototype being $BaTiO_3$. But ionic materials can also be formed with molecular ions giving a very large class of crystals like $NaNO_2$, KCN etc . . . The purpose of this book is not to give a general review of the electronic and vibrational properties of all possible defects in all possible ionic materials, but to present particularly simple, selected systems, from which it will be possible to understand more complex systems. For this reason, we shall focus essentially on point defects in alkali-halides, (thermal defects[†] – Schottky and Frenkel defects – will not be treated). In this chapter, we shall introduce principal defects which will be the subject of the following chapters. These defects, in alkali-halides, are related to the anionic or cationic sublattice or to impurities. We shall also discuss the main defects in other types of ionic solids such as alkaline earth oxides, fluorides and silver halides. Excitons, relaxed or unrelaxed, which play a fundamental role in the radiolysis of alkali-halides will be discussed in chapter 4.

[†] This subject is very well covered by the first three papers of the book "Point Defects in Solids" Vol. 1. ed. J. H. Crawford Jr. and L. M. Slifkin (Plenum Press, 1972).

1.1. Principal point defects in alkali-halides

In alkali-halides, the best known defects are connected with either halogen vacancies or interstitials; to the first category belong electron centers, to the second the hole centers.

1.1.1. Electron centers

The nomenclature commonly used in the literature is generally too complicated and therefore, in the following, we shall use the following representation: a F_n-center will be made up of n anion vacancies with n associated electrons for a F_n^+ there will be only $n - 1$ electrons, for F_n^-, $n + 1$ electrons, and so on.

1.1.1.1. The F-center.
An initially transparent KBr crystal becomes purple if exposed to high energy ionizing radiation (far ultraviolet, X-rays, gamma rays, electrons). The same coloration is found upon heating the crystal at 600°C in the presence of potassium vapors. These effects were first observed by Pohl and collaborators [3] after World War 1. At that time they suggested that the coloration was due to point defects they called color centers or F-centers (from the German Farben – color). However the same coloration may also be obtained by high temperature electrolysis of the crystal [4]. The color of these crystals so treated is due to an optical absorption band located at 620 nm, at room temperature (fig. 1.1); in this figure, the absorption coefficient [i.e. $\ln(I_0/I)/d$] is plotted versus light wavelength; I_0 and I are the

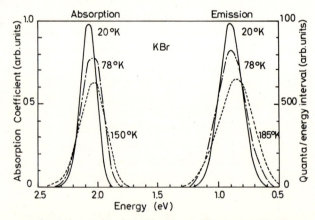

Fig. 1.1. F-center absorption and emission bands in KBr at several temperatures [from W. Gebbart and A. Kühnert, Phys. Stat. Solidi **14** (1966) 157].

incident and transmitted light intensity respectively, and d is the crystal thickness. This absorption band shifts slightly in peak position and decreases in halfwidth as the temperature is reduced. At 4 K, this band has an almost Gaussian shape, with a maximum at 2.06 eV (604 nm) and a halfwidth of 0.16 eV. Upon excitation of the colored crystal with light in the spectral range of the absorption band an emission band is observed which, at 4.2 K, is located at 0.92 eV and has a halfwidth of 0.22 eV. The absorption and emission band are separated by a *Stokes shift* of 1.14 eV. The same phenomenology is observed in most alkaki-halides, with the exception of some cesium halides for which the absorption band is structured [5].

The maximum of the F-absorption band in the various alkali-halides follows an empirical law, the so-called Mollwo–Yvey law [6]:

$$E_F = 17.7\, a^{-1.84}, \tag{1.1}$$

where E_F is the absorption band maximum, and a, the nearest anion–cation distance expressed in angströms.

The existence of broad optical bands (instead of the narrow line spectra of atomic transitions) and the Stokes shift are direct manifestations of the coupling of the defect electron with the lattice vibrations. The effect will be treated in detail in chapter 3. There we shall show why the bandwidths follow a temperature dependence described by

$$H = H_0(\coth \hbar\omega/2\,kT)^{1/2}, \tag{1.2}$$

H_0 being the halfwidth extrapolated at 0 K (fig.. 1.2)

The F-band absorption may be used to determine the defect concentra-

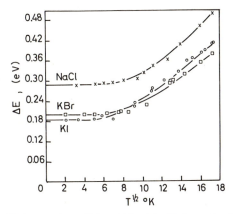

Fig. 1.2. Variation with temperature of the width of the F-absorption band at half-maximum in NaCl, KBr and KI. Full lines correspond to expression (1.2) with appropriate $\hbar\omega$ values for each crystal, [from G. A. Russel and C. C. Klick, Phys. Rev. **101** (1956)].

tion in the crystal, by considering Smakula's formula [7] which relates the defect concentration to the zero moment (i.e. the integrated area) of the absorption band:

$$N_f = 1.29 \times 10^{17} \frac{n}{(n^2 + 2)^2} \alpha_{max} H, \tag{1.3}$$

where: N is the number of F-centers per cubic centimeter, f is the oscillator strength of the optical transmission, n is the index of refraction in the appropriate spectral range, α_{max} is the absorption coefficient (in cm^{-1}) at the absorption band maximum, and H is the halfwidth of the band (expressed in cm^{-1}).

It is possible to determine the oscillator strength of the F-transition experimentally by the measuring of the F-center concentration by some other method (by EPR, the F-center being paramagnetic). In most alkali-halides this oscillator strength is found to be close to unity.

A certain number of models have been proposed for the F-center. A most interesting one, among the "wrong models", is the so-called self-trapped electron [8]. In the presence of the electron, the nearest ions will shift their equilibrium positions, thus creating a potential well which in turn will trap the electron. Such configurations called "small polarons", have been well studied [9]. However theoretical predictions about the existence of the self-trapped electron have never been verified experimentally and the F-center is made up of an electron trapped by the potential of a negative ion vacancy (fig. 1.3). The electron is in a potential well of quasi-spherical symmetry and therefore its quantum states may be described with spherical harmonics: 1s for the groundstate, 2s, 2p, etc, for the excited states. The optical transition tied to the F-band would then be due to the 1s–2p allowed transition.

This model of the F-center has been suggested by several optical measurements; the F-center luminescence is not polarized upon excitation with polarized light, which is compatible with O_h symmetry. Measurements of photoconductivity and luminescence decay time are also in agreement with such a model. These experiments will be discussed in more detail in chapter 3. However the most detailed verification of De Boer's model [10] came with EPR [11] and ENDOR measurements. The F-center is paramagnetic and for instance in KI one observes a resonance centered at $g = 1.964$ with a width of 212 G in band X. The nearly free electron value of g, is in agreement with an unpaired electron in a 1s state; the fact that g is inferior to 2 however, indicated that the groundstate is not purely 1s; this is also reasonable since the electron may not be completely localized within the vacancy and the potential is not spherical. The resonance has gaussian shape and is always wide in the alkali-halides; however in some of them (CsBr and NaF) the EPR

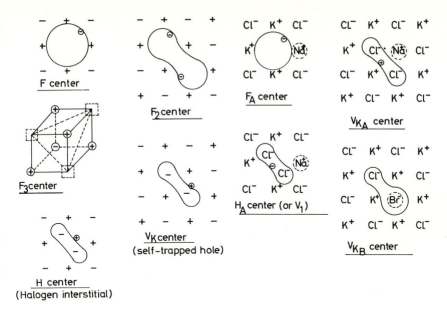

Fig. 1.3. Structure of usual color centers in alkali-halides.

spectrum is somewhat structured and the number of peaks varies with the relative orientation of the magnetic field and crystalline axes. Such a structure is due to the hyperfine interaction between the F-electron spin and the spins of the nearby ions; it is reasonable that the same interaction be responsible for the width of the resonance in the other alkali-halides, such as KI and KCl. The resonance relaxation time is very short at room temperature, but increases considerably when T is decreased and may reach values as long as tens of minutes at 1.2 K. Thus it is very easy to saturate the resonance and perform ENDOR measurements, i.e. eliminate this saturation by pumping

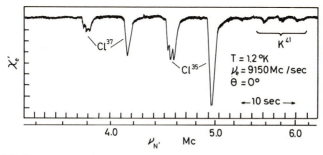

Fig. 1.4. ENDOR spectrum of F-centers in KCl: this spectrum corresponds to the hyperfine interaction of the F-center electron with nuclei of the first chlorine shell; a strong isotopic effect can be observed (from Feher [12]).

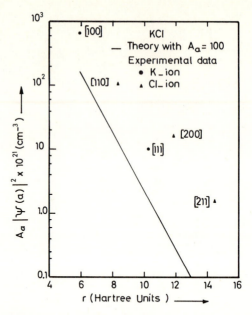

Fig. 1.5. Spatial distribution of the F-center electron wavefunction measured by ENDOR (from Holton and Blume [13]). Full line: the theoretical curve obtained with the Gourary–Adrian model (see text).

the spin system with the radiofrequency corresponding to the hyperfine interaction energy.

The first ENDOR measurements on F-centers (and among the first ENDOR measurements at all) were carried out by Feher [12] in 1957 (fig. 1.4). Holton and Blum [13] have subsequently published a detailed study in which they have shown that the isotropic part of the hyperfine interaction is proportional to the contact interaction between the F-electron and the nearest ions; thus its measurement yields a "map" of the wavefunctions of the F-center in its groundstate. The isotropic part allows the identification of the shells of neighbouring ions with which the electron interacts; the identification was carried out to the eighth-shell. In fig. 1.5 we show the spatial distribution of the grounstate wavefunction of the F-center determined by Holton and Blum[†]. Using this technique the effect on the wavefunction of uniaxial stress and electric fields has also been studied [14].

Besides the F-band, the colored crystals show higher energy bands,

[†]In this figure, the theoretical curve is that calculated by Gourary and Adrian (Solid St. Phys. 10, 1960) using a spherical potential and neglecting lattice distortions, polarization changes and exchange. In order to take these terms into account, and to obtain the electron density on a particular ion, it is necessary to multiply the envelope function $|\Phi_F|^2$ by an amplification coefficient which depends essentially only on the nature of the particular ion.

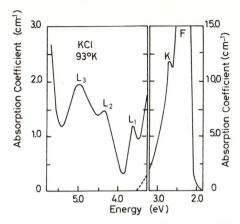

Fig. 1.6. The absorption bands connected to the F-center in KCl (from Lüty [15]).

called, in order of increasing energy, K, L_1, L_2, L_3 bands; an example is given in fig. 1.6 for KCl. The K-band has an oscillator strength of 0.1, considerably greater than that of the L-bands discovered by Lüty [15]. The K-band is thought of as originating from the transitions from the 1s groundstate to the 3p, 4p ... np levels on the high energy side of the 2p level [16].

It is interesting to note that some of these higher levels are in the conduction band of the crystal, since it is possible to observe K-band excited photocurrent even at 4.2 K. Thus the L-bands present an interesting theoretical problem since they are optical resonances to states which are degenerate with the continuum of the conduction band. Calculations based on the concept of resonant states have been performed by several authors [17, 18] and are in agreement with the experimental results. Another F-center related band is the β-band, appearing on the fundamental absorption edge. It is thought of as being due to a localized exciton transition near an F-center [19].

In conclusion, the F-center structure is now well established, its principal characteristics well known; thus this defect is an ideal subject for very detailed experimental studies. We shall use it as an example a great deal in this book.

1.1.1.2 Centers connected with a halogen vacancy. The F-center (F'- center in the literature) has been known for a long time [20] (see fig. 1.7.). Its optical properties have not been studied as much as those of the F-center because this center has an extremely wide absorption band which is located in the same spectral region as the F-band. This defect has been identified by illuminating an additively colored crystal at low temperatures ($T < 100$ K)

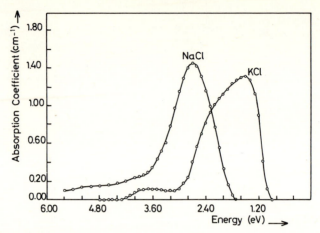

Fig. 1.7. Absorption bands of F⁻-centers in NaCl and KCl (from Lynch and Robinson [20]).

in the K-band; what is observed is a decrease of the F-band and a corresponding appearance and increase of the F⁻-band. If then the crystal is illuminated only in the F⁻-band, the F⁻-centers disappear, the F-band returns and at the same time a photocurrent may be observed. With this type of experiment Pick [21] has shown that it was necessary to destroy two F-centers to create one F⁻-center. At the low temperatures involved the vacancy mobility is practically zero and thus only electronic processes may be involved in the reactions; the experiments show that the F⁻-centers are formed with the reactions:

$$F + h\nu \rightarrow F^+ + e^-,$$
$$e^- + F \rightarrow F^-. \tag{1.4}$$

Following the first of these reactions, it is possible to observe a new absorption band, called the α-band, on the low energy side of the fundamental absorption edge. This band is similar to the β-band (in KBr the α-band is located at 6.14 eV and the β-band at 6.42 eV). With the same type of experiment used for the F⁻-center, it was possible to show that the α-band is due to the F⁺-center [19] i.e. the halogen vacancy. As for the β-band, the α-optical transition is due to localized exciton transition perturbed by the vacancy [22, 23]. It is to be noted that the α-band allows the determination of the halogen vacancy concentration in the alkali-halides.

1.1.1.3. Centers associated with two halogen vacancies. A KCl crystal irradiated at room temperature shows the F-band at 550 nm, but also another band at 820 nm called the M-band. A similar band is found in the other alkali-halides (review papers given in ref. [24]), with the absorption peak energy following a Mollwo–Yvey law [25]. The center associated with this band is the F_2-center, i.e. two F-centers in the nearest neighbour position (i.e. in the $\langle 110 \rangle$ direction, fig. 1.3). Among the many experiments giving information about the structure of the defect we shall cite two of the most significant: creation kinetics studies and the optical determinations of the defect symmetry. The creation kinetics may be studied by following the behavior of the F- and F_2-bands, the intensity of which is a measure of the relative defect concentration; in general, in very pure crystals, the concentration of F_2-centers is found to vary proportionally with the square of the F-center concentration [26]; this bimolecular-type reaction indicates that the F_2 is made up of two F-centers. The same result has been obtained by Van Doorn [27] for additively colored crystals. Furthermore the same relationship between concentrations has been observed in strongly X-irradiated crystals at low temperatures [28], where the centers are not mobile and thus the only possibility of forming an F_2-center is that an F-center be created in the nearest neighbour position of another F-center. The defect symmetry may be determined from the luminescence associated with the F_2-band excitation (emission max. at 1080 nm in KCl). An ensemble of anisotropic defects of the same type (i.e. having D_{2h} symmetry along the $\langle 110 \rangle$ axis of the crystal) will emit partially polarized luminescence if excited with polarized light. The degree of polarization will vary with the orientation and the polarization of excitation and emission light. This problem has been treated in detail by Feofilov [29] for all relevant orientations and defect symmetries. It is thus sufficient to confront experimental results with the established theory to obtain the defect symmetry. The F_2-centers do emit polarized luminescence [30] and the relative degrees of polarization for the various orientations are compatible only with an orientation of the defect along either the $\langle 110 \rangle$ or the $\langle 211 \rangle$ directions. For geometrical reasons the $\langle 110 \rangle$ configuration has been preferred. In a crystal containing F_2-centers, there is in principle an equal number of such defects in each of the six equivalent $\langle 110 \rangle$ directions, and therefore the F_2-absorption does not show any linear dichroism. It is however possible to selectively bleach the F_2-centers by illuminating with F-band light propagating in the [100] direction and polarized in the [011] direction; evidently only those centers having a component of their dipole oriented in the [011] direction will actually absorb the light. In the excited state, these centers will be destroyed and thus the crystal will become dichroic. By using this effect in a variety of experimental geometries it is possible to determine the defect symmetry, and in fact Ueta [31] has so demonstrated its $\langle 110 \rangle$ orientation. Thus the F_2-center symmetry is well established; so is its electronic structure, since the total absence of paramagnetism indicates that the two halogen vacancies are indeed decorated with two electrons [32].

More recently evidence has been found for the F_2^+- and F_2^--centers in most alkali-halides [33]. In these cases however the optical bands are not Gaussian-like; they actually feature pronounced structures both in absorption and in emission [34]. These centers and these types of structured transitions will be discussed in some detail in chapter 4. The F_2^--center has a very detailed vibrational structure in its optical bands (figs. 4.3 and 4.4); in LiF the zero-quantum line (the so-called zero-phonon line) is sufficiently narrow (1 Å at 10 000 Å) to allow detailed spectroscopic measurements (see chapter 4) [35].

The F_2-type centers have D_{2h} symmetry and thus no degenerate electronic states are allowed; thus the electronic transition associated dipole will have a definite direction: for a [110] center, there are three possible orientations, [110], [001], [110]. A crystal in which all the F_2-centers have been aligned will then be dichroic in the spectral region of the F_2-absorption. Thus Okamoto [36], and more recently Engström [37], were able to observe the higher excited states of the F_2-center, located in the region of the F- and K-bands. The resulting level scheme can be interpreted qualitatively using the simple model of an H_2 molecule embedded in a dielectric continuum [38], taking for the proton distance the vacancy separation. Evarestov [39], and Mayer and Wood [40] have given more complete theoretical treatments. Excitation in any of the excited levels of the F_2 always yields the emission corresponding to the transition from the lowest excited states ($^1\Sigma_u \rightarrow {}^1\Sigma_g$ using molecular terminology) [41]; it is thus evident that there will be non-radiative transitions to populate the emitting level $^1\Sigma_u$. In LiF [42], this result is confirmed by detection of stress induced dichroism; the absorption band shows a first moment change (for a theory of moment analysis see ref. [43]), which indicates that the external perturbation lifts the orientational degeneracy of the center. The emission band also shows a zero moment change which is only possible if there is mixing of nearby levels into the $^1\Sigma_u$ emitting level; the proximity of various levels makes non-radiative transitions among them more likely.

Upon excitation in the higher excited states with polarized light, the F_2-center can be easily oriented even at low temperatures [44] (fig. 1.8). The orientation mechanism has been clarified by Schneider [45]: the F_2-center is ionized to form F_2^+, which in turn has absorption in the same region as the F-band. The excitation light affects only a fraction of these centers which reorient during a non-radiative transition. The reoriented F_2^+-centers can then trap free electrons to form oriented F_2-centers.

Upon excitation with F-band light, a crystal containing F- and F_2-centers will show time dependent absorption bands [46] corresponding to the metastable triplet state of the F_2-center [47]. The behaviour of this triplet state is very interesting, especially for a deep understanding of the nature of non-radiative transitions and we shall discuss this subject in detail in chapter 4.

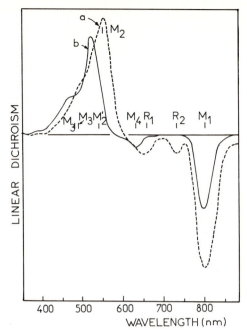

Fig. 1.8. Linear dichroism of a KCl crystal at 77 K where all F_2^--centers have been aligned along [110] (from Engstrom [34]).

1.1.1.4. More complex aggregates of the F-centers. With the methods discussed in the preceding paragraph, several authors have determined the characteristics of centers such as F_3, $F_3^{\pm}$ ([35], [49] and [24]) in the alkalihalides; these defects are made up of three halogen vacancies forming an equilateral triangle in the (111) plane and containing respectively 3, 2 or 4 electrons. These defects have C_{3v} symmetry and thus may have doubly degenerate electronic states; this makes them good systems for the study of the optical effects of the Jahn–Teller effects (see § 3.4.4.).

The more complex F-center aggregates are not well known because the relative reaction kinetics are complicated and the defects become more and more symmetric as their size increases. The large agglomerates (say with more than fifty F-centers) can be studied with small angle X-ray scattering, which yields their mean size and concentration [50] they have also been studied by light scattering [51]. Such agglomerates, also called colloids, are generally metallic globules the EPR spectrum of which it has also been possible to study [52]. In a particular case (LiF irradiated with thermal neutrons) Lambert [53] et al. have observed the formation of two-dimensional metallic agglomerates; these systems are extremely interesting for the study of a very pure metal in two dimensions.

1.1.2. Hole centers

Our knowledge is less satisfying for hole centers[†]. Actually, essentially the V_k-center (the self-trapped hole), the V_f-center (antimorph of the F-center), and the H-center (the interstitial halogen atom), are known in some detail. Many questions remain open for the other hole centers, and we shall treat them only briefly.

1.1.2.1. The V_k-center or self-trapped hole. Castner and Kanzig [54] have studied the EPR spectrum of LiF crystals irradiated with X-rays inside the microwave cavity at 77 K. Using the hyperfine structure, they have shown that the defect associated with this spectrum is simply a hole self-trapped on a pair of F^- (fluoride) ions (fig. 1.3); upon trapping the ions come closer yielding the molecular ion F_2^- (not to be confused with the homonymous center). Similar defects have been observed in KCl, NaCl and KBr, yielding respectively Cl_2^- and Br_2^- molecular ions at the appropriate lattice sites. These defects are very well known by now in most alkali-halides (see table 1.1). They show several absorption bands, of which only one corresponds to a truly allowed transition [55].

It is easy to reorient the V_k-center with polarized light excitation in the band corresponding to the allowed transition [56]; in particular the study of the subsequent thermal disorientation of the defect upon heating by measur-

Table 1.1
Reorientation temperature and absorption bands of V_K centers in several alkali-halides [55].

X_2^-	Matrix	Reorientation temperature (°C)	Wavelength of absorption bands (nm)	Relative intensities of bands	Halfwidth at half-maximum (eV)
F_2^-	LiF	−160	348	200	1.20
			750	1	
Cl_2^-	KCl	−100	365	100	0.81
			750	1	0.37
Br_2^-	KBr	−130	385	445	0.73
			705	9.5	0.26
			900	1	
I_2^-	KI	−180	400	340	0.55
			585	56	0.36
			800	1	0.22
			1150		0.19

[†]This subject is very well covered by the following review articles: M. N. Kabler in Point Defects in Ionic Crystals–op. cit.; N. Itoh–Cryst Lattice Defects 3 (1973) 115; D. Schoemaker – in Defects and their Structure in Non Metallic Solids – op. cit.

ing the optical dichroism has given information of the mechanism of hole diffusion in the alkali-halides [57]. The V_k-center has $\langle 110 \rangle$ D_{2h} symmetry and diffused by 60° jumps among the six equivalent [110] directions. This diffusion process is highly peculiar. The associated jump frequency has been measured by Keller and Murray [57] for KI:

$$f = 4.0 \times 10^{13} \exp(-0.273/kT) \, \text{S}^{-1}, \tag{1.5}$$

with kT expressed in eV units. The jump energy is 0.54 eV in KCl, and thus in this substance the hole mobility temperature is correspondingly higher.

The V_k-center also has theoretical interest since its effective mass is infinite and thus the ensuing localization makes ordinary band theory inapplicable [58]. Another important problem is to understand why the hole may self-trap and not the electron, which would form a molecular ion like K_2^+ in KCl, for example. However, such a defect has never been observed and some theoretical work has shown that it could not be stable [59].

1.1.2.2. The H-center.
Upon irradiation at 4 K a KCl crystal shows essentially two absorption bands: the F-band and another band in the near ultraviolet (336 nm) called the H-band. By illuminating the H-band region with polarized light, Compton and Klick [60] have observed defect reorientation and have shown that the center is $\langle 110 \rangle$ oriented on the basis of the kinetics of the induced dichroism. Thereafter Känzig and Woodruff [61] by EPR measurements have shown that the defect consisted of an interstitial halogen atom forming a Cl_2^- or Br_2^- molecular ion which in turn is bound to a Cl^- or Br^- ion, displaced towards it by the interaction: thus the center of the "molecule" is at the site of the original displaced ion (fig. 1.3). The discovery of this defect was of some importance, since by showing that the irradiation created a Frenkel pair it has brought a better understanding of the creation mechanism of defects under the action of ionizing radiations. We shall discuss this subject in some detail in chapter 5.

In all the alkali-halides the H-center is oriented in the $\langle 110 \rangle$ direction, with the exception of LiF, for which the orientation is $\langle 111 \rangle$ [62]. The center becomes mobile at temperatures in the range of 40 to 60 K for the various alkali-halides (see table 1.2); however it may rotate freely about its axis at lower temperatures (11 K for KCl) [63]. As for the V_k-center phonon, allowed and forbidden transitions have been observed [63]. It may be interesting to show how difficult it actually is to calculate the defect orientation. For instance Dienes et al. [64] have calculated the defect orientation for the H-center in NaCl and KCl, by taking into account the electrostatic energies, polarization and lattice distorsion effects, dipole–dipole and repulsive inter-

Table 1.2
Thermal stability of the H-center in several alkali halides.

Matrix	Measurement	Migration temperature (K)	References
LiF	EPR	60	a
	EPR	42–60	b
KCl	Optical absorption	33–44	c
KBr	EPR	20–40	b
KI	Optical absorption	55–70	d

[a] Y. H. Chu and R. C. Mieher, Phys. Rev. Letters **20** (1968) 1289.
[b] W. Känzig and T. O. Woodruff, Phys. Rev. **109** (1958) 220; J. Phys. Chem. Solids **9** (1958) 70.
[c] A. Behr, H. Peisl and W. Waidelich, Phys. Letters **24A** (1967) 379.
[d] H. N. Hersch., Phys. Rev. **148** (1966) 928; J. D. Konitzer and H. N. Hersch, J. Phys. Chem. Solids **27** (1966) 771.

actions, and found a ⟨111⟩ orientation. In order to obtain agreement with experiment the authors had to consider the hole not confined to the halide molecular ion, but also shared by the next halide ions in the ⟨110⟩ direction. Ueta [65] has measured the activation energy characterizing the bleaching of H-centers upon irradiation with electron pulses; the low values of 0.075 eV for KCl and 0.09 eV, for KBr show the high mobility of these interstitials.

1.1.2.3. The interstitial halide ion. This defect is not paramagnetic and is not associated to well defined optical absorption bands; thus its existence has been inferred from indirect information. In KBr, Itoh [66] and collaborators have observed that the F^+ absorption band would disappear in several stages in crystals irradiated at 4 K and then heated in the range 10 to 30 K. As for thermal bleaching in irradiated metals [67], they have concluded that halide ion interstitials existed which recombined with F^+-centers at temperatures which are lower the closer these interstitials are to this defect. Thermal conductivity measurements [68] in KCl and KBr, and also determinations of variations in the lattice parameters and sample length [69, 70] support the same conclusions.

More recently [71], optical properties of this defect in KBr have been studied. Two optical absorption bands around 230 nm have been observed with an energy separation of 0.31 eV. Two emission bands are present when the defect is excited in its absorption bands which corresponds, in the same way as for the self-trapped exciton emission (see chapter 4), to a singlet and a triplet initial state. All these results are interpreted assuming that the Br^- ion is in an interstitial position at a site of symmetry T_d. The groundstate is non degenerate when the excited state which corresponds to the 2p is orbitally

Table 1.3
Optical properties of the Br⁻ interstitial ion in KBr, (I-center) [71].

	Transition energy (eV)	Halfwidth (eV)
Absorption bands	5.38	0.30
	5.69	0.30
Emission bands	4.40	0.44
	2.48	0.40

degenerate, this degeneracy is lifted by the spin–orbit coupling and then yields two absorption bands (table 1.3).

1.1.2.4. Interstitial aggregates. We do not know other simple defects due to hole or interstitial halogens. For crystals irradiated at temperatures higher than 77 K, Peisl and co-workers [72, 73] have shown with measurements of density and lattice parameter variations that Frenkel pairs are created and that the interstitials do not leave the crystal; Sonder and Walton [74], by thermal conductivity have shown that interstitial aggregates are formed whose size increases with the irradiation temperature. The first stage of this aggregation is the di-interstitial observed by Itoh and Saidoh [75], which is called the V_4-center. The exact nature of this defect is not yet clear. According to Elango and Nurakhmetov [76], this defect should in KBr be a Br_3^- linear molecular ion along $\langle 100 \rangle$ in a divacancy (K^+ and Br^- vacancies): if they are right, the association of two H-centers would bring about a local reconstruction of the lattice with the formation of a cationic interstitial.

It is possible that impurities may be necessary to form aggregates at temperatures above about 50 K. Hobbs and co-workers [77], have shown by electron microscopy that halogen interstitials form dislocation loops with normals parallel to the $\langle 100 \rangle$ crystal axis, for crystals irradiated at room temperature. These loops have diameters between tens and hundreds of angströms. In their experiments they are also obliged to assume a reconstruction of the lattice during the aggregation of H-centers which forms cationic interstitials and vacancies. Such an assumption is in very good agreement with electric dipoles which are observed in relation with these H-center aggregates [78].

It is surprising that hardly any work has been performed on defects created by halogen additive coloration. The last works date from more than ten years ago and a review of such work will be found in the work of Schulman and Compton. It is certain, however, that techniques such as EPR and ENDOR, coupled to the availability of very pure crystals will make possible a better understanding of this particularly complex subject.

1.1.3. Alkaline sublattice defects

There is very little information on defects in the alkali-ion sublattice, and for a simple reason: ionizing radiation displaces only the halogen ions. In order to create defects in the alkali sublattice, the ions must be displaced by direct collision, or use thermal treatments which will create Frenkel pairs. Let us first discuss the existence of the antimorph of the F-center (a hole trapped by a positive ion vacancy).

Känzig [79] has observed such a defect by electron spin resonance in LiF irradiated at 77 K and has called it V_F-center. In this defect, the hole is not localized on the cationic vacancy but forms a V_K-center with two neighboring F^- ions; the internuclear axis of the F_2^- molecule ion lies in a {100} plane and parallel to a ⟨110⟩ direction. The molecular bond is bent through an angle of 8°. The same defect has been observed more recently in KCl [80], where it has been formed in a quite complex way. The geometry of the defect is the same as in LiF with a bending of the bond of 2.8°. In KCl, the V_F-center has a strong σ polarized transition at 362 nm and a less intense weakly π polarized transition near 750 nm; optical excitation at 77 K can cause preferential orientation of the molecule ions. Thermal reorientation of this defect occurs in the neighborhood of 220 K in LiF and 110 K in KCl. This defect has also been observed in KI and probably in NaI [81].

The antimorph of the H-center (interstitial alkali atom) has been seen in LiF [82] irradiated with electrons and neutrons at 77 K. This latter defect has the same configuration (a molecular ion Li_2^+ oriented along ⟨110⟩), as the H-center; it is very stable since it bleaches out only for temperatures higher than 400 K. It is created by direct collision, in contrast to F-type centers which are formed by excitonic recombination. Practically no work has been performed on color centers formed by additative coloration in halogen vapours since the Schulman and Compton book has been published. Main bands formed in KI, (V_2 at 290 nm and V_3 at 330 nm) seem to be related to the same defect along a ⟨110⟩ axis and could be an I_3^- molecule ion stabilized by an K^+ vacancy [83].

1.1.4. Impurity defects

There is a considerable amount of work on impurities in alkali-halides, thus in this book we shall only discuss some topics, considering only the most well known impurity systems of which we shall give the main characteristics.

1.1.4.1. Alkaline impurities. Most of the alkali-halides may be grown as mixed crystals with any content of the doping alkali metal. The most interesting results concern crystals doped in the range of 10^2 to 10^3 ppm; it is easy to see that in this case the impurities will be isolated and that it will be possible to perturb in a controlled fashion the well known color centers by associating them to one such impurity. Resonant and localized modes will also be introduced in the phonon or electron density of states of the crystal by the presence of such impurities [84].

If a doped crystal (KCl:Li for instance) is additively colored to form F-centers, some of these may be associated with the Li impurity, so that the chlorine vacancy will be surrounded by five K^+ ions and an Li^+ ion; the O_h cubic symmetry will reduce to C_{4v} and the resulting crystal field will lift the threefold degeneracy of the 2p-like excited state of the F-center, yielding two levels, $2p_z$ along the [100] direction joining the defect and the impurity and the doubly degenerate $2p_{x,y}$ levels in the perpendicular plane. Two absorption bands are effectively observed, the one of them having twice the intensity of the other (fig. 1.9). This defect, called the F_A-center has been studied in several alkali-halides; Lüty [85] has given a beautiful review of its many interesting properties.

It is possible to obtain different types of F_A-centers by varying the alkaline impurity of the host crystal. Two main categories may be identified. In category I [$F_A(I)$ centers] the emission is in the same spectral region as the

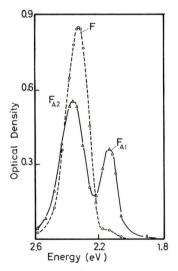

Fig. 1.9. Optical absorption of the F_A (Na^+) center in KCl at 50 K (dashed line: F-center absorption band). [from F. Lüty and W. Ziezelmann, Solid St. Commun. **2** (1964) 179].

F-emission in the same crystal. Upon excitation at $T > 60$ K with polarized light in one of the two absorption bands, the defects reorient and the crystal becomes dichroic. This process is now well understood: upon excitation with [001] polarized light corresponding to the $2p_z$ absorption, only the F_A-centers with their C_4 axis in that specific [001] direction will absorb the light; in the relaxed excited states such defects will be able to reorient in either [010] or [100] directions. This reorientation will decrease the number of centers in the [001] direction. Illuminating for a sufficiently long time it is possible to orient all the F_A-centers in the [010] and [100] directions. Upon excitation to the $2p_{x,y}$ states with [001] polarized light, the centers will be oriented in the same [001] direction. The reorientation in the relaxed excited state is thermally activated and the reorientation rate can be expressed as

$$\eta = \eta_0 \exp(-E_R/kT) \tag{1.6}$$

where η_0 is of the order of 10^3s^{-1} and E_R about 0.1 eV in most alkali-halides.

The behavior of the F_A(II) centers is very different. Their emission takes place at much lower energies than the F-center; for instance, in KCl, at low temperatures, the F-center emission is located at 1.24 eV, whereas that of the type-one F_A(Na) center is at 1.12 eV and that of type-two F_A(Li) at 0.46 eV. The F_A(II) center reorients easily in the relaxed excited state (RES) and the rate is temperature independent. Fritz and Lüty [85] have proposed a particularly simple and elegant model to interpret these results, whose basic feature is the radical difference between the RES for the type I and II centers. In the first case the potential seen by the electron is not too

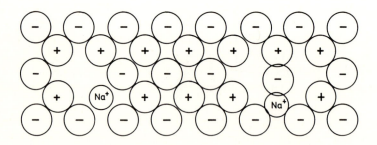

Fig. 1.10. Relaxed excited state configurations of the two kinds of F_A-centers, (from Lüty [85]).

different from the corresponding F-center, the vacancy will remain fixed at low temperature and the defect will not reorient. In the second case, one of the ions nearest neighbors of the vacancy will move to occupy the empty slot between the impurity and a positive ion (fig. 1.10), thus the defect will reorient automatically in the RES, loosing the memory of its initial orientation; its electron will find itself in a potential which is different from that of the F-electron in the corresponding RES, and the emission will reflect this difference both in spectral position and decay times. This model is also confirmed by considerations of the ionic radii of the ions involved.

A crystal of KCl:Na irradiated at 77 K shows an absorption band called the H_A-band [86] (V_1 in the older literature) which is due to an interstitial halogen atom stabilized by the Na impurity; this was demonstrated by Delbecq and co-workers [87] with EPR measurements. The interstitial is more stable than the H-center; in fact, in LiF the H-center bleaches out at 60 K [88] whereas the H_A(Na) is not affected until 110 K [89]. It is interesting to note that it is easy to convert optically H_A-centers into H-centers at 4 K, i.e. to untrap the interstitials to put them back in a pure area of the lattice [90].

The hole may also be stabilized by impurities, as in the case of NaF:Li [91]. The relative center becomes mobile at higher temperatures than the isolated hole.

Finally other defects have been observed, such as the F_{2A}-center, i.e. and the F_2-center nearby an alkali impurity [92].

1.1.4.2. Halogen impurities. The halogen interstitial is stabilized in crystals such as KCl, KBr and KI doped with fluorine, forming respectively molecular ions such as FCl^-, FBr^- and FI^-, which occupy a negative ion vacancy and are oriented in the ⟨111⟩ direction [93]. Above 150 K these defects bleach yielding a Cl^- (or Br^-, I^-) in a normal lattice position and the F-interstitial which may migrate. Analogously the hole (V_k-center) may be stabilized by these impurities; in KCl:I it forms the molecular ion ICl^- oriented in the ⟨110⟩ direction [94]. The equivalent of the F_A-center has been observed by Pelsers and Jacobs [95] in CsBr:Cl; in this case also, the F-band is split by the lifting of the 2p threefold degeneracy. A large number of such defects has been determined and it would take too long to give a complete review of these defects.

1.1.4.3. Hydrogen. It is easy to introduce in the crystal the substitutional H^- ion during the crystal growth, or by heating the additively colored crystal in hydrogen atmosphere. The H atoms will diffuse and be trapped by F-centers to form H^- which are called U-centers [96]. These defects, non-paramagnetic, give origin to an absorption band in the ultraviolet

(192 nm in NaCl and 228 nm in KBr for instance). They are also associated with well studied vibrational localized modes, which we shall discuss in the following chapter. By photochemical reactions at various temperatures it has been possible to produce the U_1-center, i.e. the H^- ion in an interstitial position (1/4, 1/4, 1/4); the U_2-center, i.e. the interstitial H atom and the U_3-center, the substitutional H atom [98].

Finally, most of these impurities give origin to localized exciton bands on the low energy side of the crystal fundamental absorption [97]. In KCl:I for instance, the I ion yields bands at 6.7 eV and 7.2 eV at 77 K, whereas the first exciton peak is at 7.9 eV. As discussed by Fowler the emission processes of these centers are complex and are not interpreted at this time.

1.1.4.4. Divalent metals.   A great deal of work has been done on divalent impurities (e.g. Ca in NaCl); (see the review articles of ref. [99] for details). In non-irradiated crystals, the impurity is associated to a positive ion vacancy, forming an electric dipole in the $\langle 110 \rangle$ direction; in some cases the vacancy is in the next nearest neighbour position and the dipole is oriented in the $\langle 100 \rangle$ direction. The dipole is thermally dissociated and free dissociation energies and vacancy migration energies can be obtained from ionic conductivity measurements, ref. [100]. It is also possible to study the thermal reorientation of these dipoles by studying the dielectric losses or, in a more elegant fashion by using the ionic thermocurrents method developed by Bucci and Fieschi [101]. This latter method works as follows: the dipoles are oriented at room temperature by an applied electric field; they are then frozen in their configuration by rapidly cooling the sample. The temperature is then allowed to increase in zero electric field; at well defined temperatures these dipoles reorient and yield depolarization currents (the ionic thermocurrents) from which it is possible to measure the reorientation energy of the dipoles (this phenomenon is formally similar to thermoluminescence). As for the monovalent impurities, it is possible to associate irradiation defects to these impurities. An F-center associated with such a dipole form the F_{1z}-center [102] (Z_1 in the old nomenclature). The nature of the other centers (Z_2, Z_3) has not been clarified yet. At sufficiently high temperatures some positive ion vacancies dissociate thermally from the impurity; and can be frozen-out by quenching the crystal. Irradiation at 77 K creates then Frenkel pairs and the neutral interstitials are trapped by these vacancies to form X_2 molecules, with an X ion nearby in a $\langle 100 \rangle$ direction [103].

1.1.4.5. Heavy metals.   The most studied metallic impurities correspond to s^2 ions, such as Tl, In, Ga, Sn, Pb. In fact these impurities, and particularly the thallium impurity center, have been the first such systems to be understood quantitatively. Very early Seitz [104] proposed a model for the spectrum of the Tl center in which the optical bands were attributed to the electronic transitions between the atomic states of the impurity itself, the surrounding lattice acting only as a perturbation. Although the Seitz model and subsequent modifications (see chapters 3 and 4) do give a good qualitative explanation for the optical spectra, many features cannot be explained on the basis of such a simple model and in fact only very recently a reasonably quantitative explanation of absorption and emission spectra has been achieved. In its groundstate the Tl (and similar) ions have two 6s electrons; in the first excited state one of the two electrons is in a 6p state. As we shall see in chapter 3, such a configuration leads to four levels, with allowed transitions for two of them. These transitions lead to the A and C absorption bands in the 200 to 300 nm spectral region. In between them lies the B-band, with an oscillator strength which depends strongly on temperature: this band is due to a forbidden transition, which is partially allowed by lattice vibrations. Although a one-to-one correspondence exists between atomic levels and impurity bands, these latter ones are very broad (about 0.2 eV width). This indicates a strong electron–lattice interaction. This coupling constitutes one of the most interesting aspects of studies on the Tl and similar centers. These centers also have a characteristic luminescence which can be excited in any of their absorption bands or by irradiation of the crystal with ionizing radiation (including the near UV) or particles. In fact this luminescence is at the basis of the well known scintillation properties of NaI:Tl.

A great number of metallic impurities may be introduced into the alkali-halides. An important one is the silver impurity, since it can exist in many ionization states [105]: Ag^+, Ag^- and Ag°. Ag^- is particularly interesting since it is isoelectronic with Tl but has opposite charge: thus in the crystal this impurity is surrounded by cations, whereas Tl^+ is surrounded by the more polarizable anions. Ag°, having only one electron, is in some ways similar to the F-center. Finally Ag^+ allows a detailed study of forbidden transitions since all of its optical bands are weak and highly temperature dependent. The Cu^+-center has the same type of interest, although its characteristics are very different from those of Ag^+. In particular, in most hosts the Cu^+ ion is possibly in an off-center position [106]. This hypothesis is also confirmed by the existence of resonant vibrational mode associated with Cu in the far infrared [107].

This "off-center" position of ions in crystals is a very important and fundamental phenomenon which will be discussed in more detail in the next chapter. However let us present first clear evidence of such an effect [108]. Optical absorption bands can be observed from substitutional Ag^+ in alkali-halides corresponding to transitions from the groundstate (configuration d^{10}: ^{1}s) to excited states (configuration d^9s: ^{1}D or ^{3}D); these transitions are parity forbidden (and spin forbidden for the second excited state) and are partly allowed by the mixing of the ^{1}D or ^{3}D state with higher odd states by odd phonons. If the impurity is centered at an inversion symmetry point, the oscillator strength of the absorption increases with temperature according to the following classical law:

$$f = f_0 \coth(\hbar\omega/2kT), \tag{1.7}$$

where $\hbar\omega$ is the mean energy of the odd phonon inducing the transition. Such behavior is effectively observed in KCl:Ag for example, but not at all in CsBr:Ag, RbCl:Ag and RbBr:Ag. In these last three cases, the oscillator strength of the transition decreases when the temperature increases. This effect can be understood assuming that the Ag^+ has an ionic radius which is small enough to force it to be in "off-center" positions in the cation vacancy. Then the ion is no longer at an inversion symmetry point of the potential and the crystal field will mix odd and even parity levels allowing the transitions at low temperature. If the temperature increases, the ion is moving more and more in the vacancy and will spend longer and longer times at the center of the vacancy which causes a decrease of the oscillator strength.

1.1.4.6. Rare earth ions.   Rare earth ions in alkali-halides show complex absorption and emission bands. The ions in their divalent state are associated with a positive ion vacancy in the ⟨110⟩ direction, and thus the local

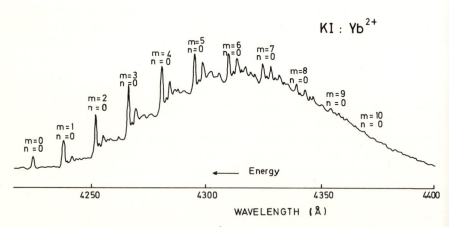

Fig. 1.11. Low energy emission band of Yb^{2+} in KI at 10 K, (from Wagner and Bron [110]).

symmetry is C_{2v}. Bron and Heller [109] have studied the rich absorption and emission spectra of the Sm^{+2} in several alkali-halides; in order to interpret the experimental data, they had to assume that the ion and the vacancy are slightly displaced toward each other. The coupling with lattice vibrations causes well pronounced vibration structures as can be seen in the case of the lowest energy emission band of $KI:Yb^{+2}$[110] (see fig. 1.11).

1.1.4.7. Anionic and molecular impurities. Anionic impurities may be introduced by several methods in the alkali-halides, either as atomic (for instance O^- or S^-) or molecular ions [111, 112]. These latter defects are interesting since they may only be studied in certain cases as impurities in a crystal matrix, when they are unstable or even non-existent as free ions. These molecules have, in general, lower symmetry than that of the ion they substitute in the lattice, and therefore there exist several equivalent equilibrium positions for them, separated by a barrier; this barrier may be overcome by tunnelling (in the case of O_2^- and OH^- for instance) [113] or by thermal activation.

Polyatomic molecules with a center of symmetry may yield "para-elastic" centers, which can all be oriented in a particular direction by applying the appropriate uniaxial stress: Känzig [114] was the first to observe this effect in $KCl:O_2^-$. If no center of symmetry is present these defects may also be para-electric, and thus be oriented by an electric field (for instance OH^- in KCl). We shall discuss these effects in more detail in chapter 2.

In general the absorption–emission spectra of these defects are very complicated: in fig. 3.19 we show the absorption spectrum of NO_2^- in KCl. In fact, the optical spectra reflect not only the vibration–rotation spectrum of the molecules themselves, but also the phonon spectra (including possible localized modes) of the host crystal. Thus by optical absorption spectroscopy it is possible to obtain detailed information on the vibrational dynamics of these systems.

1.2. Color centers in other ionic solids

There are many types of ionic solids where color centers have been found. We shall limit our review to three specific ones: the alkali-earth fluorides and oxides, and the silver halides. This choice does indicate that our understanding of the phenomenology in these systems is reasonably satisfactory; however the reader should not infer that color centers are limited to these solids; in fact some interesting results have been found in molecular ionic crystals such as $NaNO_3$ or in the perovskites, such as $KMgF_3$.

1.2.1. The alkali-earth fluorides†

These substances form two groups: those with cubic crystal symmetry (CaF_2, SrF_2, BaF_2), in which each fluorine is surrounded by a tetrahedron of positive ions (T_d symmetry), and those with quadratic symmetry, in which each fluorine is located at a site of C_{2v} symmetry (cassiterite structure, MgF_2). Here we shall only consider the first group of crystals [115].

Upon additive coloration of CaF_2 in calcium vapours, two absorption bands at 370 and 520 nm appear. The second band has been assigned to the F_2-center, whereas the first one is connected with both F_2- and F-centers. In fact in these substances the F-centers have a marked tendency to aggregate rapidly, even if the crystals are rapidly quenched. As in the alkali-halides the F_2-center has D_{2h} symmetry, but this time around a $\langle 100 \rangle$ direction. It is also possible to orient these centers by excitation in the F-band with polarized light at 77 K. The centers reorient thermally at about 270 K, with an activation energy of 0.98 eV. The F_3-center has been found optically [116]: it is made up of three F-centers aligned along a $\langle 100 \rangle$ direction.

The V_k-center has been observed by EPR: it may be formed by X-irradiation in CaF_2 crystals doped with rare earth ions: these ions must be added to trap electrons. The defect has D_{2h} symmetry, with the twofold axis in the $\langle 100 \rangle$ direction. It is connected with an optical absorption band centered at 320 nm and here again irradiation with polarized light will orient it totally in one specific direction.

The H-center has been observed in crystals submitted to heavy irradiation with 1 MeV electrons at 20 K. It appears that in such materials, H-centers are always stabilized by impurities. The efficiency of coloration by ionizing radiation is very small, but it is now quite well established that Frenkel pairs can be formed by such radiations as in alkali-halides [118]. As we shall see in chapter 4, the exciton in alkali-halides become very quickly self-trapped; it is the same for alkaline earth-fluorides. In CaF_2 for example the bandgap is equal to 12.2 eV whereas luminescence of the lowest triplet state of the self-trapped exciton occurs at 4.43 eV [119].

Finally, it is well known that CaF_2 is a universally used matrix for rare earth ions [120], yielding crystals with photochromic properties [121], or laser media. However, even though crystal field effects have been thoroughly studied, less attention has been paid to the effects of electron–photon coupling in these systems.

Substitutional hydrogen has also been introduced in CaF_2: it yields no optical absorption up to 200 nm, but shows an infrared absorption which is

† For a good review, see W. Hayes [115].

interesting because it is due to a localized mode in T_d symmetry, whereas the analogous mode in the alkali-halides was in O_h symmetry. Practically, the known impurities in these crystals are limited to these two types.

1.2.2. Alkali-earth oxides [122, 123]

Color centers in these substances are not well known since these materials cannot be easily obtained as pure monocrystals. Since they are grown by cathode discharge, it is practically impossible to eliminate impurities such as transition metal ions (Ni, Fe etc). The study of these impurities in such hosts is, on the other hand, well developed and, in particular, crystal field effects have been investigated. Contrary to alkali-halides, the oxides are not colored by ionizing radiations: more precisely, the radiations only displace electrons and do not create Frenkel pairs. Point defects may be produced by additive coloration (in the metal vapour), by electrolytic coloration or by irradiation with massive particles.

1.2.2.1. Defects connected with charge changes.

In alkali or alkali-earth halides, the hole self-traps and becomes mobile only above a certain temperature. Things are different in the alkali-earth oxides: ionizing radiation creates electron–hole pairs; the electrons will then be trapped by impurities which will change their state of charge, and the hole will be trapped at defects; these defects are generally positive ion vacancies sometimes associated with impurities. For isolated vacancies, the hole will localize on one of the six nearby oxygens, forming thus a center with tetragonal symmetry along a $\langle 100 \rangle$ axis. This defect, the V^--center (a positive ion vacancy with a charge compensating hole) has been observed by EPR in MgO [124]. In this crystal, the defect originates a broad optical absorption band (width about 1 eV) located at 2.3 eV [125]. It is not possible to orient this defect optically; this fact indicates that the hole jumps from one oxygen to the other even at 4 K. Impurities may perturb the V^--center; for instance a proton which may be trapped inside the vacancy in the $\langle 100 \rangle$ direction $\langle 126 \rangle$, a fluoride ion which substitutes an O_2^- [127] or even a trivalent metal ion.

1.2.2.2. Defects associated with oxygen vacancies.

By additive coloration it is possible to create centers with one vacancy, two vacancies etc., just as in the alkali-halides. The centers associated with one oxygen vacancy, are known to have two different charge states, yielding the F^+- and F-centers respectively with one and two electrons in the vacancy.

Table 1.4
F- and F^+-absorption and emission bands in some alkaline earth oxides.

Matrix	F-center		F^+-center	
	Absorption (eV)	Emission (eV)	Absorption (eV)	Emission (eV)
MgO	4.95 [a,b]	3.1 [b,c]	5.03 [b]	2.4 [b]
CaO	3.70 [a,d]	3.353 [g]	3.1 [d]	2.058 [g]
SrO	3.0 [a,e,f,h]		2.5 [e]	
BaO	2.0 [a,b]			

[a] K. C. To and B. Henderson, J. Phys. C: Solid St. Phys. **4** (1971) L216.
[b] L. A. Kappers, R. L. Kros and E. B. Hersley, Phys. Rev. **B1** (1970) 4151.
[c] Y. Chen, J. L. Kolopus and W. A. Sibley, Phys. Rev. **186** (1969) 465.
[d] W. C. Ward and E. B. Hensley, Phys. Rev. **175** (1968) 1230.
[e] B. P. Johnson and E. B. Hensley, Phys. Rev. **180** (1969) 931.
[f] R. Bessent, B. C. Cavenett and I. C. Hunter, J. Phys. Chem. Solids **29** (1968) 1523.
[g] B. Henderson, S. E. Stokowski and T. C. Ensign, Phys. Rev. **183** (1969) 826.
[h] F. A. Modine, Phys. Letters **A34** (1971) 413.

The F^+-center. This paramagnetic defect has been identified first by EPR [128] in MgO. Contrary to the F-center in KCl, the electron is strongly localized in the oxygen vacancy and the only hyperfine interaction detectable is with the nearest neighbours. The defect is associated with an optical absorption band in the different alkali-earth oxides (cf. table 1.4). In CaO, the absorption band has vibrational structure and a zero-phonon line at 355.7 nm: this fact makes the F^+-center in CaO an ideal system for the study of the Jahn–Teller effect in a triply degenerate electronic state [129, 130].

The F^+-center perturbed by a cation vacancy or a surface. In a MgO crystal irradiated with fast neutrons, the number of F^+-centers decreases rapidly upon heating. This decrease is due to the coagulation of the centers with cation vacancies created by the irradiation [131]. EPR measurements show that the largest hyperfine interaction is with the nearest neighbour Mg ion in the ⟨100⟩ direction [132]: this means that the electron is in some fashion repelled by the positive ion vacancy and therefore localizes on the oxygen vacancy.

A powder sample of MgO irradiated with neutrons or γ-rays in vacuum shows a purple coloration [133]. This color disappears quickly if the powder is placed in an oxygen atmosphere. The defects responsible for the coloration are shown by EPR to possess C_{4v} symmetry. It seems then that these defects are the F^+-center or the surface of the crystal†; the defect disappears

---

† For surface point defects, see: A. M. Stoneham in: Defects and their Structure in Non-Metallic Solids, op. cit. p. 355.

upon exposure to oxygen due to adsorption of a O_2 molecule which then would diffuse to the vacancy to form O_2^-. This model seems to be corroborated by more detailed studies by Tench [134] with EPR: the temperature variation of the hyperfine constants is much greater for these centers than for the bulk F^+-centers. This effect can be easily understood since the hyperfine interaction is mainly with the ions nearest neighbours of the vacancy and its temperature variation will be greater the larger the vibrational amplitudes of the ions; such amplitudes will be larger for surface ions than for bulk ones. More studies should be performed on this center. In particular in CaO, where the emission has vibrational structure, one could obtain interesting results on the vibrational modes at the surface of the crystal.

The F-center. This center is less easy to study since it is not paramagnetic: $F^+ \rightarrow F$ photoconversion measurements were necessary to identify it. Henderson and co-workers [135] have shown that in CaO there is an emission originating from the triple 3p state to the 1s groundstate; thus its decay time is particularly long ($\tau \approx 3 \times 10^{-3}$ s at 4.2 K). The non-radiative transition between the ^{1}p state reached in absorption and the ^{3}p emitting state is very likely caused by crossing of the levels during the relaxation of the defect after the optical absorption. This effect has been used by Merle d'Aubigne and co-workers to study the magnetic circular dichroism in the emission of the F-center in CaO [136].

Other color centers have been identified in the alkali-earth oxides, particularly F-center aggregates similar to the analogous centers in the alkali-halides [137]. This subject is too large to be discussed here in any detail.

1.2.3. Point defects in silver halides

It is curious that we have comparatively little information on substances that have great industrial importance such as the silver halides (see for example the review papers [138]). Although they are very sensitive to visible and ultraviolet radiation at room temperature, they are not colored by ionizing radiation; in AgCl and AgBr, the light creates small aggregates of interstitial Ag° [139] which the developing agent will increase in size (amplification 10^9) to yield the blackening of a photographic plate [140]. The absence of coloration has discouraged physicists, who then turned to the study of intrinsic excitons, particularly easy in these materials with a gap of 2.5 to 3.5 eV. The high static dielectric constant makes these compounds well suited for polaron studies: an electron in the conduction band polarizes the surrounding crystal and this polarization follows the electron in its displacements giving it a large effective mass. The effect may be described microscopically as the coupling of the electron to longitudinal optical phonons. The point defects studies in these crystals have been mainly impurities which yield localized excitons; one of the best examples is AgBr:I⁻, whose very interesting properties have not been completely interpreted yet [141].

Research in color centers in silver halides has been stimulated by the theoretical work of Buimistrov on the energy levels of the F-center [142]. As for semiconductors, he finds that the electronic wavefunction radius should be large compared to the vacancy dimension. This justifies a continuum model (see chapter 4) for the center: the slow electron interacts strongly with the lattice and such a system can be considered as a bound polaron. The calculation predicts a weak dissociation energy for the F-center (0.0192 eV in AgBr); the optical transitions associated with this center must then be located in the far infrared where no experimentalist has ever looked for color centers.

1.2.3.1. The bound polaron in silver halides. Brandt and Brown [143] have effectively observed an IR absorption at low temperatures in crystals of AgCl and AgBr illuminated with light in the spectral region of band-to-band transitions (fig. 1.12). For AgBr this absorption consists of a very narrow line at 168 cm⁻¹ and of a group of lines at higher energy which correspond to one, two-phonon transitions, etc. These lines – of which only the first is depicted in fig. 1.12 – are regularly spaced, and the spacing of 124 cm⁻¹ corresponds to the frequency of the LO phonon of the crystal (at the zone

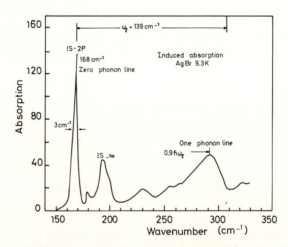

Fig. 1.12. Optical absorption of bound polarons in AgBr created by illuminating the crystal in its fundamental absorption, (from Brandt and Brown [143]).

center). This absorption corresponds to a bound polaron since at these temperatures the hole is not mobile in silver halides [144]; in fact the radiation creates electron–hole pairs, the holes self-trap and the electrons are trapped by some impurity, around which they will gravitate. Since the orbit radius will certainly be much larger than the trap size, the absorption spectrum will not depend on the detailed nature of the impurity: one may use an hydrogenoid model of the type used by Buimistrov [142]. The zero-phonon line corresponds then to the 1s–2p transition and the absorption band at $198 \, \text{cm}^{-1}$ to excitation into the continuum, i.e. the bound polaron dissociation. More sophisticated calculations interpret the observed absorption spectra completely [145]. The model has been verified by Zeeman effect on the narrow line at $168 \, \text{cm}^{-1}$ [146]. These measurements allow the determination of the m_z mass of the polaron in the 2p state:

$$E_{p,+1} - E_{p,-1} = heH/M_z c, \tag{1.8}$$

where $E_{p,M}$ is the energy of the p-state with orbital angular momentum M, in a magnetic field H. The bound polaron model yields a value for the mass of $0.320 \, \text{eV} \pm 0.007$ a.u.

What then is the nature of the defect X around which the polaron revolves? A first series of investigations indicates that it is not a bromine vacancy; such an F-center created at 4 K would be stable; instead, the absorption induced by pumping in the band-to-band region decreases, when pumping is stopped, in some tens of seconds. This decrease may be described by eq. (2.10) in which x is the number of bound polarons [143];

$$t = 10^4 \exp(-13.7x) + \exp(6.4x)/17, \tag{1.9}$$

the above expression is valid for time intervals between 10 and 1000 s. If we retain only the first term in eq. (1.9), we find:

$$dx/dt = -\exp(13.7x)/137\,000. \tag{1.10}$$

Thus x decreases logarithmically and non-exponentially, i.e. the decrease is not due to a monomolecular process for it does not correspond to an intrinsic process of the bound polaron. On the other band the decrease does not vary with temperature (in the measured range 9–50 K). A de-excitation model which interprets these results may be the following: the bound polarons extend reasonably far from the X^+ impurity so that there may be some overlap of the 1s wavefunctions of near polarons. If one such interaction dissociates (or destroys) the bound polaron, then the rate of diminution of polarons will decrease as their concentration decreases. Since the wavefunctions decrease exponentially with distance, the destruction rate will decrease exponentially as a function of mean distance between bound polarons; distance which is proportional to the number of polarons: this is what is actually observed.

1.2.3.2. Localized electrons in silver halides.   These polarons are a kind of localization of electrons in the crystal. Kanzaki and Sakuragi [147] have shown that at 4 K electrons are localized at silver interstitial ions or substitutional divalent cation impurities. Around 20 K, these shallow-localized electrons become thermally unstable and the absorption due to "deep-localized" electrons around 0.5 to 1 eV tends to dominate the spectrum. From several experiments, it appears that diatomic molecular states such as Ag_2 or Ag_2^+ are responsible for the deep localized electrons. At higher temperature above 50 K, clustering of localized electrons proceeds further and results in a broad absorption around 1.0 to 2.0 eV. The excitation induced absorption remains stable after termination of excitation. The origin of these stable localized electrons above 50 K can be assigned as molecular clusters such as Ag_3 or Ag_4 which presumably correspond to the photographic latent image.

1.2.3.3. The self-trapped hole and the self-trapped exciton in silver halides.   In crystals containing electron traps, Höhne and Stasiw [144] have observed, at low temperatures, the EPR spectrum of the Ag^{2+} ion after irradiation with band-to-band phonons. The EPR spectrum is exactly the same for the different electron trapping impurities, which indicates that the defect produced is not associated with any particular impurity. This result is very important because it shows that, in silver halides, the hole self-traps on a Ag^+ ion, forming an Ag^{2+} ion at a normal lattice site. The hole becomes unstable above 50 K, but we have no knowledge of the mechanism of thermal bleaching.

An optical absorption band associated with the self-trapped hole has been observed at 9200 cm^{-1} in AgCl and 7100 cm^{-1} in AgBr, by several authors [148, 149]. This band would correspond to a charge transfer transition between Ag^{2+} and a nearby halogen. Upon irradiation in this band, the self-trapped holes bleach out and the electron-traps charge changes: thus this irradiation would displace the hole which would then annihilate on the traps.

The self-trapping of excitons is, in principle, less probable in crystals more covalent than alkali-halides. However Hayes et al. [150] have observed in AgCl under irradiation with X-rays at 4 K, a broad emission band peaking at 2.52 eV. This luminescence shows magnetic circular polarization with a resonance in a low magnetic field. It is then possible to detect optically the electron spin resonance of the excited state of the defect giving rise to the emission (see § 4.3.1). From these measurements they conclude that the self-trapped exciton is the source of the emission and that it involves the 4d wave function of the silver ion which becomes an Ag^{2+} ion in a normal site. In AgCl crystals containing bromine substitutional impurities an additional center emits in the 2.52 eV range and excited state EPR shows that the hole is self-trapped on a silver ion close to the impurity.

1.2.3.4. Bose–Einstein condensation of free excitons and bi-excitons in AgBr.
At the end of this chapter, let us just present briefly the very interesting results obtained by Hulin and his co-workers [151] from which they claim to have observed the Bose–Einstein condensation of free excitons, i.e. the formation of an electron–hole liquid in AgBr in equilibrium with excitons and bi-excitons. This condensation is a very well known phenomena; for example, it is responsible of the superfluid transition of He^4. Such a condensation is expected from excitons which can be considered as good bosons if the Bohr radius of the exciton is small compared to the mean distance between excitons.

AgBr has an indirect bandgap with the conduction band minimum at the

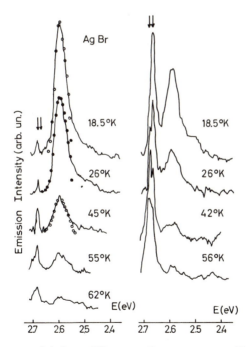

Fig. 1.13. Luminescence of AgBr at different cooling temperatures. The pulsed excitation intensity is $\approx$ 50 kW cm^{-2} on the left (a), and $\approx$ 5 MW cm^{-2} on the right (b). The cirlces show the calculated lineshape for electron–hole liquid recombination at effective temperatures of 50 K (T = 18.5 K), 70 K (T = 26 K) and 97 K (T = 45 K). The two arrows indicate the positions of free excitons X and excitonic molecules XX. (from Hulin et al. [151]).

Γ-point and the valence band maxima at the L-point. The indirect gap has an energy of 2.70 eV, while the direct gap is about 5.0 eV. For pure material, radiative band-to-band and free exciton recombination in AgBr is allowed by symmetry with the cooperation of acoustical phonons. The emission spectrum changes drastically with excitation intensity as well as with temperature as is shown in fig. 1.13. In this figure, three main features are observed: the free exciton recombination line X at 2.68 eV, a broad emission band at 2.6 eV which corresponds to the electron–hole liquid and a bi-exciton line XX at 2.67 eV close to the line X. The broad emission is attributed to the TA_L-phonon assisted annihilation of electron–hole pairs in the condensed liquid phase. Excellent agreement between the observed emission and the calculated lineshape for electron–hole liquid recombination can be obtained provided that one introduces an effective temperature T_L for the liquid higher than the average sample temperature T (fig. 1.13). This assignment is confirmed by the existence of a threshold for the appearance of this emission with the intensity of the excitation light and with temperature.

In this material, electron–hole liquid can coexist with molecular bi-excitons as is shown in fig. 1.13. This point is discussed in more detail by Hulin et al. [151] where they propose a phase diagram (temperature versus density of excitons).

References to chapter 1

[1] M. Yuste, L. Taurel, M. Rahmani and D. Lemoyne, J. Phys. Chem. Solids **37** (1976) 961.
[2] R. J. De Jongh and A. R. Miedema, Adv. Phys. **23** (1974) 1.
[3] R. W. Pohl, Proc. Phys. Soc. **49** (1937) fasc. 3; Z. Phys. **39** (1938) 36.
[4] J. H. Schulman and W. D. Compton, Color Centers in Solids (Pergamon Press, Oxford, 1963); J. J. Markhman, F-centre in alkali-halides, Solid St. Phys. suppl. 8 (1966).
[5] T. A. Fulton and D. G. Fitchen, Phys. Rev. **179** (1969) 846.
[6] E. Mollwo, Nachr. Ges. Wissen, Göttingen (1931) p. 97; H. Ivey, Phys. Rev. **72** (1947) 341.
[7] A. Smakula, Z. Phys. **59** (1930) 603.
[8] J. H. Schulman and W. D. Compton, Color Centers in Solids (op. cit: see refs. [22] p. 135).
[9] See refs. [124–127] (p. 138 of ref. [4]).
[10] J. H. De Boer, Recueil Trav. Chim. Pays Bas, **56** (1937) 301.
[11] H. Seidel, in: physics of color Centers, ed. by W. B. Fowler (Academic Press, New York 1968).
[12] G. Feher, Phys. Rev. **114** (1959) 1219, 1245.
[13] W. C. Holton and H. Blum, Phys. Rev. **125** (1962) 89.
[14] C. E. Bailey, J. Phys. Chem. Sol. **31** (1970) 2229; J. F. Reichert and P. S. Perschan, Phys. Rev. Letters **15** (1965) 780; Z. Usmani and J. F. Reichert, Phys. Rev. **180** (1969); 482 Z. Usmani and J. F. Reichert, Phys. Rev. B1 (1970) 2078.
[15] F. Lüty, Z. Phys. **160** (1960) 1.
[16] D. Y. Smith and G. Spinolo, Phys. Rev. **140** (1965) A2121; G. Iadonisi and G. Preziosi, Nuovo Cim. **B48** (1967) 92.

[17] Physics of Color Centers (op. cit.) p. 91.
[18] K. Kojima, Int. Symp. on Color Centers (Urbana, 1965); F. Bassani, Int. Symp. on Colour Centers (Urbana, 1965).
[19] C. J. Delbecq, P. Pringshein and P. H. Yuster, J. Chem. Phys. **19** (1951) 574.
[20] D. W. Lynch and D. A. Robinson, Phys. Rev. **174** (1968) 1050.
[21] H. Pick, Nuovo Cim. VII (Ser X) no. 2 (1958) 498.
[22] F. Porret and F. Lüty, Phys. Rev. Letters **26** (1971) 843.
[23] F. Bassani and N. Inchawspe, Phys. Rev. **105** (1957) 819. T. Timusk, J. Phys. Chem. Sol. **26** (1965) 849.
[24] H. Rabin and W. D. Compton, Solid St. Phys. Vol. **16** (1966); W. Von Der Osten, Defects and their Structure in Non-Metallic Solids, ed. B. Henderson and A. E. Hughes, (Plenum Press, New York, 1976) p. 237.
[25] H. Ivey, Phys. Rev. **72** (1947) 341.
[26] W. D. Compton and H. Rabin, Phys. Rev. Letters **7** (1961) 57.
[27] C. Z. Van Doorn, Phys. Rev. Letters **4** (1960) 236.
[28] E. Sonder and W. A. Sibley, Phys. Stat. Solidi **10** (1965) 99.
[29] P. P. Feofilov, Dokl. Akad. Nauk. **92** (1953) 545, 743.
[30] J. Lambe and W. D. Compton, Phys. Rev. **106** (1957) 684.
[31] M. Ueta, J. Phys. Soc. Japan **7** (1952) 107.
[32] W. E. Bron, Phys. Rev. **128** (1962) 627; H. Blum, Phys. Rev. **125** (1962) 509.
[33] I. Schneider and H. Rabin, Phys. Rev. **140** (1965) 1983.
[34] Y. Farge, G. Toulouse and M. Lambert, C.R. Acad. Sci. **262** (1966) 1012.
[35] Y. Farge and M. Fontana, Solid St. Commun. **10** (1972) 333; D. B. Fitchen, H. R. Fetterman and C. B. Pierce, Solid St. Commun. **4** (1966) 205.
[36] F. Okamoto, Phys. Rev. **124** (1961) 1090.
[37] A. Engström, Phys. Rev. **B11** (1975) 1657.
[38] R. Hermann, M. C. Wallis and R. F. Wallis, Phys. Rev. **103** (1956) 87.
[39] R. A. Evarestov, Opt. and Spectrosc. **16** (1964) 198.
[40] A. Meyer and R. F. Wood, Phys. Rev. **133A** (1964) 1436.
[41] C. J. Delbecq, Phys. Rev. **171** (1963) 560.
[42] M. Maki, Y. Farge and M. P. Fontana, Int. Conf. on Color Centers (Reading, 1971).
[43] C. H. Henry and C. P. Slichter, in: Physics of Color Centers. (op. cit.)
[44] T. J. Turner, R. De Batist and Y. Haven, Phys. Stat. Sol. **11** (1965) 267, 535.
[45] I. Schneider, Phys. Rev. Letters **24** (1970) 1296.
[46] I. Schneider and M. E. Caspari, J. Appl. Phys. **33** (1962) 3387.
[47] H. Seidel, Phys. Letters **A7** (1963) 27; Yu. V. Fedotov, N. N. Bagmut and I. V. Matyash, Sov. Phys. Sol. St. **13** (1972) 519.
[48] Y. Farge, Czech. J. Phys. **B20** (1970) 611.
[49] L. F. Stiles and B. D. Fitchen, Phys. Rev. Letters **17** (1966) 689; J. A. Davis and B. D. Fitchen, Solid St. Commun. **7** (1969) 1363; M. Inoue, J. Phys. **48** (1970) 1694; L. Norenzo, E. Reguzzoni and G. Samoggia, Phys. St. Sol. **38** (1970) 369; L. Nosenzo, Phys. Letters **A32** (1970) 415.
[50] P. Durand, Y. Farge and M. Lambert, J. Phys. Chem. Sol. **30** (1969) 1353.
[51] See J. H. Schulman and W. D. Compton (op. cit.) ch. 9.
[52] C. Taupin, J. Phys. Chem. Sol. **28** (1967) 41.
[53] M. Lambert, C. Mazieres and A. Guinier, J. Phys. Chem. Sol. **18** (1961) 129; C. Taupin and D. Taupin, C.R. Acad. Sci. **264** (1967) 581.
[54] T. G. Castner and W. Känzig, J. Phys. Chem. Sol. **3** (1957) 178.
[55] C. J. Delbecq, W. Hayes and P. H Yuster, Phys. Rev. **121** (1961) 1043.
[56] C. J. Delbecq, B. Smaller and P. H. Yuster, Phys. Rev. **111** (1958) 1235.
[57] F. J. Keller and R. B. Murray, Phys. Rev. Letters **15** (1965) 198; F. J. Keller, R. B. Murray, M. M. Abraham and R. A. Wecks, Phys. Rev. **154** (1967) 812.
[58] A. N. Jette, T. L. Gilbert and T. P. Das, Phys. Rev. **184** (1969) 884; D. Ikenberry, A. N. Jette and T. P. Das, Phys. Rev. **B1** (1970) 2785; K. S. Song, J. Phys. Soc. Japan. **26** (1969)

1131; K. S. Song, J. Phys. Chem. Sol. **31** (1970) 1389; R. Gazzineli and H. G. Reik, Sol. St. Commun **8** (1970) 745.

[59] J. J. Markham, Sol. St. Phys. Suppl. 8 (1960) p. 289.

[60] W. D. Compton and C. C. Klick, Phys. Rev. **110** (1958) 349.

[61] W. Känzig and T. O. Woodruff, J. Phys. Chem. Sol. **9** (1958) 70.

[62] Y. H. Chu and R. L. Mieher, Phys. Rev. Letters **20** (1968) 1289.

[63] C. J. Delbecq, J. L. Kolopus, E. L. Yasaitis and F. H. Yuster, Phys. Rev. **154** (1967) 866.

[64] G. J. Dienes, R. D. Hatcher and R. Smoluchowski, Phys. Rev. **157** (1967) 692.

[65] M. Ueta, J. Phys. Soc. Japan. **23** (1967) 1265.

[66] N. Itoh, B. S. H. Royce and R. Smoluchowski, Phys. Rev. **137**A (1965) 1010.

[67] Y. Quere, Defauts ponctuels dans les métaux (Masson, Paris, 1967) p. 194.

[68] W. Gebhardt, J. Phys. Chem. Sol. **23** (1962) 1123.

[69] W. Hertz, H. Peisl and W. Waidelich, Phys. Letters **25A** (1967) 403.

[70] R. Balzer, H. Peisl and W. Waidelich, Phys. Stat. Solidi **27K** (1968) 165.

[71] M. Takahashii, M. Saidoh and N. Itoh, Phys. Stat. Solidi **B57** (1973) 749.

[72] R. Balzer, H. Peisl and W. Waidelich, Phys. Letters **27A** (1968) 31.

[73] R. Balzer, H. Peisl and W. Waidelich, Phys. Stat. Solidi **15** (1966) 495; Phys. Stat. Solidi **28** (1968) 207; Phys. Rev. Letters **17** (1966) 1129; Z. Phys. **204** (1967) 405.

[74] E. Sonder and D. Walton, Phys. Letters **25A** (1967) 222.

[75] N. Itoh and H. Saidoh, Phys. Stat. Solidi **33** (1969) 649; M. Saidoh and P. D. Townsend, Int. Conf. Defects in Insulating Crystals (Gatlinburg, 1977) p. 365.

[76] A. A. Elango and T. N. Nurakhmetov, Phys. Stat. Solidi **B78** (1976) 529.

[77] L. W. Hobbs, A. E. Hughes and D. Pooley, Phys. Rev. Letters **28** (1972) 234; L. W. Hobbs, in: Defects and their Structure in Non-Metallic Solids. (op. cit.).

[78] J. P. Stott and J. H. Crawford Jr, Phys. Rev. **B6** (1972) 4660.

[79] W. Känzig, Phys. Rev. Letters **4** (1960) 117.

[80] C. J. Delbecq, D. Schoemaker and P. H. Yuster, Phys. Rev. **B9** (1974) 1913.

[81] M. P. Fontana and W. J. Van Sciver, Phys. Stat. Solidi **37** (1970) 375.

[82] Y. Farge, Phys. Rev. **B1** (1970) 4797.

[83] A. Okuda, T. Harami, T. Okada, Y. Tanaka and O. Takei, J. Phys. Soc. Japan. **43** (1977) 993.

[84] See chapter 2.

[85] F. Lüty, in: Physics of Color Centers (op. cit.).

[86] R. Castner, P. Pringsheim and P. H. Yuster, J. Chem. Phys. **18** (1950) 887, 1564.

[87] J. L. Kolopus, C. J. Delbecq, D. Schoemaker and P. H. Yuster, Bull. Am. Phys. Soc. **12** (1967) 467; C. J. Delbecq, E. Hutchinson, D. Schoemaker, E. L. Yasaitis and P. H. Yuster, Phys. Rev. **187** (1969) 1103; F. W. Patten and F. J. Keller, Phys. Rev. **187** (1969) 1120.

[88] Y. H. Chu and R. L. Mieher, Phys. Rev. **188** (1969) 1311.

[89] D. F. Daly and R. L. Mieher, Phys. Rev. Letters **19** (1967) 637.

[90] D. Schoemaker, Phys. Rev. **B2** (1970) 1148.

[91] I. L. Bass and R. L. Mieher, Phys. Rev. **175** (1968) 421.

[92] I. Schneider, Phys. Rev. Letters **16** (1966) 743.

[93] D. Schoemaker, Phys. Rev. **149** (1966) 693; Phys. Rev. **B3** (1971) 3516.

[94] L. S. Goldberg and M. L. Meistrich, Phys. Rev. **172** (1968) 877; C. J. Delbecq, D. Schoemaker and P. H. Yuster, Phys. Rev. **B3** (1971) 473.

[95] L. Pelsers and G. Jacobs, Phys. Rev. **188** (1969) 1324.

[96] J. H. Schulman and W. D. Compton (op. cit.) p. 163; W. B. Fowler (op. cit.) p. 123; W. Hayes and J. W. Hodby, Proc. R. Soc. **A294** (1966) 1438.

[97] W. B. Fowler, Color Centers in Solids (op. cit.) p. 164 and following.

[98] M. de Souza, A. D. Gongora, M. Aegerter and F. Lüty, Phys. Rev. Letters **25** (1970) 1426; M. de Souza and F. Lüty, Phys. Rev. **B8** (1973) 5866.

[99] N. Nagasawa, J. Phys. Soc. Japan. **27** (1969) 1535; S. Radhakrisna and B. V. R. Chowdari, Phys. St. Solidi (a)**14** (1972) 11; L. W. Barr and A. B. Lidiard, Physical

Chemistry – An advance treatise, vol. 10 (Academic Press, New York, 1970) p. 151; M. Hartmanova, Phys. Stat. Solidi (a)7 (1971) 303; V. K. Jain, Phys. Stat. Solidi (b)44 (1971) 11.

[100] Y. Adda and J. Philibert, La diffusion dans les solides (Bibliothèque des Sciences et Techniques Nucléaires, INSTN and PUF, 1966).

[101] G. Bucci and R. Fieschi, Phys. Rev. Letters 12 (1964) 16.

[102] F. Rosenberger and F. Lüty, Sol. St. Commun. 7 (1969) 249.

[103] Y. Farge, J. Phys. Chem. Sol. 30 (1969) 1375; M. Ikeya, N. Itoh, T. Okada and T. Suita, J. Phys. Soc. Japan 21 (1966) 1304.

[104] F. Seitz, J. Chem. Phys. 6 (1938) 150.

[105] W. Duetz, Phys. Stat. Solidi 34 (1969) 95; W. Kleeman, Z. Phys. 234 (1970) 362; K. Kojima, J. Phys. Soc. Japan. 28 (1970) 1227; M. Saidoh, N. Itoh, and M. Ikeya, J. Phys. Soc. Japan. 25 (1968) 1197; V. Topa, Rev. Roum. Phys. 12 (1967) 781.

[106] R. Sittig, Phys. Stat. Sol. 34K (1969) 189; S. A. Mach and W. J. Van Sciver, Phys. St. Solidi 46 (1971) 193.

[107] K. H. Timmesfeld, Phys. Letters A32 (1970) 385.

[108] T. Matsuyama, M. Saidoh and N. Itoh, J. Phys. Soc. Japan 39 (1975) 1486 and refs. therein.

[109] W. E. Bron and W. R. Heller, Phys. Rev. 136A (1964) 1433.

[110] M. Wagner and W. E. Bron, Phys. Rev. 139A (1965) 223; Phys. Rev. 139A (1965) 233.

[111] J. R. Brailsford, J. R. Morton and L. E. Vannotti, J. Chem. Phys. 49 (1968) 2237.

[112] J. R. Morton and L. E. Vannotti, Phys. Rev. 174 (1968) 448.

[113] V. Narayanamurti and R. O. Pohl, Rev. Mod. Phys. 42 (1970) 201.

[114] W. Känzig, J. Phys. Chem. Sol. 23 (1962) 479.

[115] W. Hayes, in: Non-Metallic Crystals, ed. S. C. Jain and L. T. Chadderton (Gordon and Breach, 1970); Crystals with Fluorite Structure (Clarendon Press, Oxford, 1974).

[116] J. H. Beaumont, A. L. Harmer, W. Hayes and A. R. L. Spray, Int. Conf. Color Centers (Reading, 1971) Abstract C 45).

[117] W. Hayes, R. F. Lambourn and J. P. Stott, J. Phys. (:Solid St. Phys.) 7 (1974) 2429.

[118] W. Hayes and R. F. Lambourn, Phys. Stat. Solidi (b)57 (1973) 693).

[119] P. J. Call, W. Hayes and M. N. Kabler, J. Phys. C: Solid St. Phys. 8 (1975) L60.

[120] R. Newman, Adv. Phys. 20 (1971) 197; G. Di Bartolo, Optical Interactions in Solids. (J. Wiley, New York, 1968).

[121] Z. Kiss, Phys. Today 23 (1970) 42.

[122] B. Henderson and J. E. Wert, Adv. Phys. 17 (1968) 749; A. E. Hughes and B. Henderson, in: Point Defects in Solids, ed. J. H. Crawford and L. M. Slifkin (Plenum Press, New York, 1972); B. Henderson, Defects in Crystalline Solids (Arnold, London, 1972).

[123] Defects and Their Structure in Non-Metallic Solids, ed. B. Henderson and A. E. Hughes (Plenum Press, New York, 1976).

[124] J. E. Wertz, P. Auzins, J. H. E. Griffiths and J. W. Orton, Discuss. Faraday Soc. 28 (1959) 136.

[125] Y. Chen and W. A. Sibley, Phys. Rev. 154 (1967) 842.

[126] P. W. Kirkin, P. Auzins and J. E. Wertz, J. Phys. Chem. Sol. 26 (1965) 1067.

[127] J. E. Wertz and P. Auzins, Phys. Rev. 139A (1965) 1645.

[128] J. E. Wertz, P. Auzins, P. Weeks and R. H. Silsbee, Phys. Rev. 106 (1957) 484.

[129] A. E. Hughes, J. Phys. C: Solid St. Phys. 3 (1970) 627.

[130] Y. Merle d'Aubigne and H. Roussel, Phys. Rev. B3 (1971) 1421.

[131] J. E. Wertz, J. W. Orton and P. Auzins, Discuss. Faraday Soc. 30 (1961) 40.

[132] K. C. To, A. M. Stoneham and B. Henderson, Phys. Rev. 181 (1969) 1237.

[133] R. L. Nelson and A. J. Tench, J. Chem. Phys. 40 (1964) 2736; R. L. Nelson, A. J. Tench and B. J. Harmsworth, Trans. Far. Soc. 63 (1967) 1427; A. J. Tench and R. L. Nelson, J. Colloid Interface Sci. 26 (1968) 364.

[134] A. J. Trench, Surface Sci. 25 (1971) 625.

[135] B. Henderson, S. E. Stokowski and T. C. Ensign, Phys. Rev. **183** (1969) 826.
[136] J. Duran, Y. Merle d'Aubigne and R. Romestain, Int. Conf. Color Centers (Reading, 1971) Abstract 10.
[137] Y. Chen, R. T. Williams and W. A. Sibley, Phys. Rev. **182** (1969) 960.
[138] H. Kansaki and S. Sakuragi, Photogr. Sci and Eng. **17** (1973) 69; L. Slifkin, Radiation Damage Processes in Materials ed. C.H.S. Dupuy (Noordhoff-Leyden, 1975) p. 405.
[139] F. C. Brown and N. Wainfan, Phys. Rev. **105** (1957) 93.
[140] J. H. Schulman and W. D. Comtpon, Color Centers in Solids (op. cit.) p. 270.
[141] R. K. Ahrenkiel, Sol. St. Commun. 6 (1968) 741; F. Moser, R. K. Ahrenkiel and S. L. Lyu, Phys. Rev. 161 (1967) 897; M. Tsukakoshi and H. Kanzaki, J. Phys. Soc. Japan. **30** (1971) 1423.
[142] V. M. Buimistrov, Sov. Phys. Sol. St. **5** (1963) 2387.
[143] R. C. Brandt and F. C. Brown, Phys. Rev. **181** (1969) 1241.
[144] M. Höhne and M. Stasiw, Phys. St. Solid **25K** (1968) 55; Phys. St. Solidi 28 (1968) 247.
[145] J. Devreese, Sol. St. Commun. 7 (1969) 767; K. K. Bajaj and T. D. Clark, Sol. Commun. **8** (1970) 1419.
[146] R. C. Brandt, Phys. Rev. Letters **23** (1969) 240.
[147] S. Sakuragi and H. Kanzaki, Phys. Rev. Letters **38** (1977) 1302.
[148] W. Ulrici, Phys. St. Sol. **40** (1970) 557.
[149] H. Kanzaki, Sol. St. Commun. **9** (1971) 1667.
[150] W. Hayes, I. B. Owen and P. J. Walker, J. Phys. C: Solid St. Phys. **10** (1977) 1751.
[151] D. Hulin, A. Mysyrowicz, M. Combescot, I. Pelant, C. Benoit á. Laguillaume, Phys. Rev. Letters **39** (1977) 1169.

2 | VIBRATIONAL PERTURBATIONS IN IONIC SOLIDS

In a periodic three-dimensional structure, when the harmonic approximation holds – i.e. at low temperature – the lattice vibrations may be described
by normal modes, with associated quasi-particles having an energy, a
wavevector and polarization – the phonons. This is a well known result in
solid state physics, and thus we shall briefly summarize the main hypotheses
and approximations made. What is of interest here is what happens when
a point defect is introduced in the lattice. One obvious consequence is the
breaking of translational and sometimes local symmetry: thus the vibrations
involving the impurity will not propagate, or rather, will be more or less
strongly damped spatially. Furthermore, the force constants between the
defect and its neighbors may be different from those of the intrinsic lattice
and the defect mass may be different from the ion it replaces: thus we have
either a defect-associated vibrational frequency which is outside the range
of normal mode frequencies of the host lattice and then we shall have a
localized or gap mode, or else the frequency will be degenerate with the
normal mode frequency distribution, and in that case the amplitude of
vibrations at that frequency will be amplified (resonant mode).

In some materials, a very low energy resonant mode appears which
originates from tunnel states of defects having different possible orientations (low symmetry molecule for example) or different possible stable
positions in the lattice ("off-center" impurity). These effects give rise to
interesting phenomena like para-electricity and para-elasticity that we
discuss at the end of this chapter. They can also be used to generate and
detect phonons in crystals.

Like the preceding chapter, we are giving examples here of simple
materials like alkali halides or alkaline-earth fluorides, but the general concept which result from these "simple" systems can be and have been very
useful in a lot of more complex materials that we shall not discuss here.

2.1. The dynamics of perfect crystals [1]

The theory of phonons in a perfect and finite crystal is based on two basic approximations, which we shall discuss in some detail, the adiabatic or Born–Oppenheimer approximation, and the harmonic approximation; the first is especially fundamental because it allows the separation of the electronic and vibrational dynamics. In order then to calculate the phonon dispersion curves, it is also necessary to know precisely the mutual ionic interactions inside the crystal. Several models have been used which in some cases have yielded excellent results.

2.1.1. The approximations

The crystal Hamiltonian may be written:

$$H = -\sum_l \frac{P_l^2}{2M_l} - \sum_i \frac{p_i^2}{2m} + V(\mathbf{u}, \mathbf{r}) + G(\mathbf{u}), \tag{2.1}$$

where $\mathbf{u}_j$ is the atomic displacement in unit cell j relative to its equilibrium position, and $\mathbf{r}_i$ the position of the ith electron. The first two terms correspond to the nuclear and electron kinetic energy, the third to the electron potential energy when the nuclei are displaced from their equilibrium position, and the last term yields the nuclear potential energy for a displacement $\mathbf{u}$, relative to their equilibrium position.

2.1.1.1. The adiabatic approximation. The simultaneous solution of the electron and nuclear dynamics is a difficult problem which may be considerably simplified by the intuitively reasonable Born–Oppenheimer approximation, according to which the valence electrons on a given ion follow adiabatically the displacements of this ion: during such movements, the electron does not change state, it is its own state that is progressively deformed by the nuclear movements. This approximation is very reasonable, since the very light electrons have much less inertia than the nuclei and thus can follow "instantaneously" the ionic movements. In the adiabatic approximation then the crystal wavefunction may be written as:

$$\Psi = \phi(\mathbf{u}, \mathbf{r})\, \Phi(\mathbf{u}), \tag{2.2}$$

where the function $\phi(\mathbf{u}, \mathbf{r})$ describes the ensemble of the electrons in the lattice frozen in the configuration described by the ensemble of displacements $\mathbf{u}$.

$$\left(\sum_i -\frac{p_i^2}{2m} + V(\mathbf{u}, \mathbf{r}) \right) \phi(\mathbf{u}, \mathbf{r}) = E_e(\mathbf{u})\phi(\mathbf{u}, \mathbf{r}). \tag{2.3}$$

The energy of the electrons thus depends only parametrically on the ionic positions. Upon solving the crystal eq. (2.3) with the function (2.2), we obtain:

$$H\Psi = -\sum_l \frac{P_l^2}{2M_l} \Psi + E_e(\mathbf{u})\Psi + G(\mathbf{u})\Psi. \tag{2.4}$$

or:

$$H\Psi = \phi(\mathbf{u}, \mathbf{r}) \left(-\sum_l \frac{P_l^2}{2M_l} + E_e(\mathbf{u}) + G(\mathbf{u}) \right) \Phi(\mathbf{u})$$
$$+ \left(-\sum_l \frac{\hbar^2}{M_l} \frac{\partial \Phi}{\partial \mathbf{u}_l} \frac{\partial \phi}{\partial \mathbf{u}_l} + \sum_l \frac{\hbar^2}{2M_l} \Phi \frac{\partial^2 \phi}{\partial \mathbf{u}_l^2} \right). \tag{2.5}$$

The adiabatic approximation is valid if the second term in (2.5) is negligible relative to the first term. The energies corresponding to the first and second terms of the non-adiabatic part are in fact negligible in most cases. The first term alone yields integrals of the type:

$$-\sum_l \frac{\hbar^2}{M_l} \frac{\partial \Phi}{\partial \mathbf{u}_l} \left\langle \psi \Big| \frac{\partial \psi}{\partial \mathbf{u}_l} \right\rangle = -\sum_l \frac{\hbar^2}{M_l} \frac{\partial \Phi}{\partial \mathbf{u}_l} \left[\frac{1}{2} \frac{\partial}{\partial \mathbf{u}_l} \langle \psi | \psi \rangle \right] = 0. \tag{2.6}$$

If the first term is zero, the second is very small. In the case of tight binding, the electrons follow the movements of a well defined ion, so that we can write:

$$\phi(\mathbf{u}_l, \mathbf{r}_i) = \phi(\mathbf{r}_i - \mathbf{u}_l) \tag{2.7}$$

$$\int \phi^* \frac{\hbar^2}{2M_l} \frac{\partial^2 \phi}{\partial \mathbf{u}_i^2} \, d\mathbf{r} = -\int \phi^* \frac{\hbar^2}{2M_l} \frac{\partial^2 \phi}{\partial \mathbf{r}_i^2} \, d\mathbf{r}$$
$$= -\frac{m}{M_l} \int \phi^* \frac{\hbar^2}{2m} \frac{\partial^2 \phi}{\partial \mathbf{r}_i^2} \, d\mathbf{r}. \tag{2.8}$$

This last term is just m/M_l times the electron kinetic energy, which is negligible due to the low value of the mass ratio (10^{-4}–10^{-5}).

Equation (2.5) for the (u) function can be written as:

$$\left(-\sum_l \frac{P_l^2}{2M_l} + U(\mathbf{u}) \right) \Phi(\mathbf{u}) = E\Phi(\mathbf{u}), \tag{2.9}$$

where now $U(\mathbf{u})$ is an effective potential given by $E_e(\mathbf{u}) + G(\mathbf{u})$.

2.1.1.2. The harmonic approximation. It is possible to develop the effective potential $U(\mathbf{u})$ as a function of the displacement $\mathbf{u}_i$:

$$U(\mathbf{u}) = U(0) + \sum_{i,\alpha} \frac{\partial u}{\partial u_{i\alpha}} u_{i\alpha} + \frac{1}{2} \sum_{i,\alpha} \sum_{j,\beta} \frac{\partial^2 u}{\partial u_{i\alpha} \partial u_{j\beta}} u_{i\alpha} u_{j\beta} + \cdots, \tag{2.10}$$

where the first two terms are identically zero since u is the equilibrium configuration. The harmonic approximation consists in neglecting all terms in the expansion of order higher than two. It is evident that this approximation will be that much more valuable as the displacements are small, and thus particularly at low temperatures, where anharmonic effects such as thermal expansion or phonon–phonon scattering are negligible.

2.1.2. Vibrations in a lattice with one atom per unit cell

In the harmonic approximation, eq. (2.10) becomes:

$$U = \frac{1}{2} \sum_{i,\alpha} \sum_{j,\beta} \Phi_{i\alpha,j\beta} u_{i\alpha} u_{j\beta} \tag{2.11}$$

with

$$\Phi_{i\alpha j\beta} = \partial^2 U / \partial u_{i\alpha} \partial u_{j\beta}. \tag{2.12}$$

The matrix defined in (2.12) is real and symmetric. Each of its electrons has the full lattice symmetry; the force acting on the ith atom in the α direction will be:

$$F_{i\alpha} = -\frac{\partial U}{\partial u_{i\alpha}} = -\sum_{j\beta} \Phi_{i\alpha,j\beta} u_{j\beta}, \tag{2.13}$$

so that the equation of motion will be:

$$M \frac{d^2 u_{i\alpha}}{dt^2} = -\sum_{j\beta} \Phi_{i\alpha,j\beta} u_{j\beta}. \tag{2.14}$$

The lattice periodicity allows planewave solutions:

$$\mathbf{u}_i(t) = \mathbf{v}(q) \exp[-i(\omega t - \mathbf{q} \cdot \mathbf{R}_i)], \tag{2.15}$$

$\mathbf{v}(\mathbf{q})$ being the wave polarization vector. Substituting (2.15) into (2.14), we have:

$$\omega^2(\mathbf{q}) v_\alpha(\mathbf{q}) = \frac{1}{M} \sum_{j\beta} \Phi_{i\alpha,j\beta} \exp[-i\mathbf{q}(\mathbf{R}_i - \mathbf{R}_j)] \, v_\beta(\mathbf{q})$$
$$= \sum_\beta G_{\alpha\beta}(\mathbf{q}) v_\beta(\mathbf{q}), \tag{2.16}$$

with

$$G_{\alpha\beta}(\mathbf{q}) = \frac{1}{M} \sum_j \Phi_{i\alpha,j\beta} \exp[-i\mathbf{q}(\mathbf{R}_i - \mathbf{R}_j)]. \tag{2.17}$$

The frequencies $\omega(\mathbf{q})$ may be obtained from the secular equation:

$$\mathrm{Det}[\,G_{\alpha\beta}(\mathbf{q}) - \omega^2(q)\delta_{\alpha\beta}\,] = 0. \tag{2.18}$$

The matrix G is hermitian, so that it has three real positive eigenvalues for each q: $\omega_\lambda^2(\mathbf{q})$. In a finite crystal with N units cells each containing one atom there are $3N$ modes and thus q values; it is well known that the Hamiltonian can be put in the form of second quantization:

$$H = \sum_{q\lambda}^{3N} \hbar\omega_{q\lambda} \left(b_q^{\lambda+} b_q^\lambda + \frac{1}{2} \right), \tag{2.19}$$

this Hamiltonian corresponds to the sum over $3N$ independent harmonic oscillators in wavevector space. The normal mode of frequency $\omega_{q\lambda}$ has then an energy of $(n_q^\lambda + \frac{1}{2})\hbar\omega_q$ and the quantum $\hbar\omega_q$ which propagates in the crystal with quasi-momentum $\hbar q$ is the phonon. The total vibrational energy will be given by:

$$E = \sum_{q\lambda}^{3N} \left(n_q^\lambda + \frac{1}{2} \right) \hbar\omega_{q\lambda}. \tag{2.20}$$

In the case where there is more than one atom per unit cell (J atoms), there will be $3J$ dispersion branches in wavevector space: three acoustic branches and $3J - 3$ optical branches.

2.1.3. The measurement and calculation of dispersion curves; the phonon density of states

In most problems, it is necessary to know the dispersion curves obtainable from eq. (2.18). Such curves can be obtained experimentally by inelastic neutron scattering and can then be compared with theoretical predictions based on various models. The measurements have now been performed for most of the alkali-halides, starting with NaI and KBr [2] (Fig. 2.1). We shall now briefly summarize the main characteristics of the models, considering especially those aspects which shall also be used in studying the vibrational properties of defects. The solution of (2.18) reduces to the determination of the $3\,NJ \times 3\,NJ$ matrix followed by the diagonalization of the $G_{\alpha\beta}(\mathbf{q})$ matrix for all values of q in the reciprocal lattice. Thus formulated the problem is obviously impossible to solve; we must try to simplify the matrices as much as possible, and calculate $G_{\alpha\beta}(\mathbf{q})$ only for some representative q values in specific and important directions in the reciprocal lattice. It is evident that the matrix Φ has more zero elements the shorter the range of the interactions in the crystal ($\Phi_{\alpha i,\beta j}$ is different from zero only if the atoms i and j are sufficiently near). In an ionic crystal [3] the electrostatic forces are long range; in general then, one divides the crystal around a given ion into two parts; one inner part wherein the ion perceives the displacements of the ions contained in it, and an external part where the potential may be considered static. Moreover, to calculate the matrix Φ it is necessary to use a model describing the inter-ionic interactions. We shall now discuss some of the models most frequently used, since they are also used to calculate vibrational properties of point defects.

The rigid ion model [4]. The ions are considered as points of mass M_+ and M_- respectively; the repulsive potential is only between near neighbors and only has two adjustable parameters. The equilibrium conditions determine one of the parameters, and the other may be determined empirically by compressibility measurements. In fig. 2.1, we show a comparison between measured and calculated dispersion curves with this model. Recalling that the theory uses only one parameter, determined by macroscopic measurements and not from neutron measurements, the agreement is rather good.

Models which consider the electron cores deformation. The rigid ion model neglects completely the electron core deformations which are responsible for the polarizability and the behavior of the dielectric function at high frequencies. The most simple deformation model associates to each ion a charge and an induced dipole moment without inertia. On this basis two

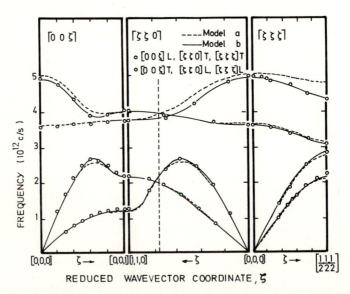

Fig. 2.1. Phonon dispersion curves in KBr measured by inelastic neutron scattering at 90 K in directions $(0\,0\,\rho)$, $(\rho\,\rho\,0)$ and $(\rho\,\rho\,\rho)$. Dashed lines represent theoretical results using a point ion model and full lines (model b) theoretical results from the shell model (from Woods et al. [2]).

mathematical models have been developed. In Hardy's model [5], the dipoles appear only when the ions are displaced from their equilibrium position; this model has been used to predict a great number of dispersion curves in alkali-halide. The other model is the shell model proposed by Dick and Overhauser [6]. In this model, the ions are assumed to be made of two parts; a hard and heavy core, and a light peripherical shell connected to the valence electrons. The core and the shell are then coupled to each other by isotropic elastic forces. Dispersion curves have been calculated for NaI and KBr [7] with the further approximations that only the negative ion is polarizable and the short range interaction is between the positive ion and the shell of the negative ion. The theory then has three free parameters which may be determined from measurements of the elastic constant C_{11} the $\varepsilon(0)$ and $\varepsilon(\infty)$ dielectric constants. Comparison with experiment is shown by the continuous lines in fig. 2.1; it is seen that the agreement is better than for the rigid ion model. Other more refined models, such as the breathing shell model, have been elaborated; of course the relative number of parameters increases. The philosophy underlying the higher theoretical complexity is that it is simply not physical to consider the ions as rigid charged spheres; the ions are soft and their mutual interactions are intrinsically complex, and simple approximations of them may just not be good enough. This situation resembles the discussion on models to determine *a priori* crystal structures.

The phonon density of states From the complete knowledge of the dispersion curves, it is theoretically easy to calculate the density $n(E)dE$, i.e. the number of vibrational states with an energy between E and $E + dE$. Unfortunately, inelastic neutron scattering measurements are long and expensive, so that the dispersion curves are determined only for some preferred directions in reciprocal space ($\langle 100 \rangle$, $\langle 110 \rangle$, $\langle 111 \rangle$). It is then necessary to extrapolate to the whole reciprocal space; in order to do so, a refined model had been devised (following the breathing shell model) in which the parameters are determined by a continuous adjusting of the theoretical results with the experimental ones; the model then yields the dispersion curves in any direction. Thus phonon density of states have been determined for several alkali-halides; among them many feature an energy gap between the acoustic and the optical modes: NaI, NaBr, KI, KBr, LiCl, and CsF.

2.2. The crystals containing point defects

An impurity absorption band was first observed in the infrared by Schafer [7]; the absorption was due to the U-center; the frequency of the vibrational mode was much higher than the maximum of the lattice frequency distribution, thus the mode is a localized mode. Thereafter a great number of defect associated modes have been detected, particularly in the alkali-halides, and review articles [8], summer schools [9] or congresses [10], have been entirely dedicated to the subject.

2.2.1. *The linear diatomic chain containing one impurity*

Using this simple example, treated in detail by Mazer and his co-workers [11] and later by Sievers [12], we can better understand the various types of defect-associated modes. An impurity of mass M' replaces either a light atom M_1 or a heavy atom M_2 of the diatomic chain, without changing the previous force constants. If M' replaces M_2 and $M' \ll M_2$ there appear two new modes, a localized mode and a gap mode (fig. 2.2). In the case that $M' \gg M_2$, then no mode appears. The reason for this difference is easy to understand. At the Brillouin zone edge, for acoustic modes, the heavy atom M_2 vibrates out of phase whereas the light atom M_1 does not move; the inverse holds for the optical modes, i.e. it is the light atom M_1 that vibrates out of phase whereas the heavy atom is immobile. Let us first consider the case $M' \ll M_2$, starting from $M' = M_2$ and then diminishing M'. At first, the impurity vibrates out of phase with neighboring M_2 atoms. As the mass decreases, the number of atoms M_2 which participate in the vibration decreases, and the mode shifts in the gap. At the center of the Brillouin zone, the mode pulls out of the dispersion branch, since here M_1 and M_2 vibrate out of phase and

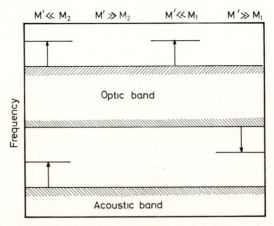

Fig. 2.2. Localized and quasi-localized modes which can appear in a linear chain of atoms containing a foreign atom (from Sievers [12]).

a reduction in M' means an increase in the frequency of the mode. When $M' \ll M_2$, the frequency of the zone-edge acoustic mode will decrease. If M' replaces the light atom, the resulting modes may be interpreted in a similar way. The displacement amplitude of the impurity atom and of the other nearby atoms have been calculated by Genzel et al. [13], and are shown in fig. 2.3. For a localized mode, the pattern of motion is similar to that of the zone-center optical branch; the amplitude however has a different spatial variation, in that it decreases exponentially. The same type of exponential decrease also takes place for gap modes (fig. 2.3). Here, similar ions vibrate out of phase, as zone-edge modes do; the electric dipole associated with these gap modes is smaller than that associated with localized modes.

If there is also a force constant change upon introduction of the impurity, localized infrared inactive modes, in which the impurity itself does not move, may appear. Resonant modes may result if the force constants are considerably reduced. Such modes had already been predicted for a mass defect in a three-dimensional mono-atomic lattice [14]. Contrary to localized gap modes, the resonant mode extends far into the lattice (fig. 2.4). The atoms

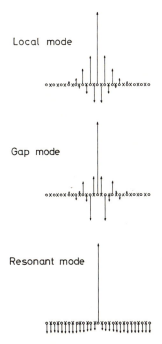

Fig. 2.3. Vibrational amplitude of neighboring atoms of the impurity atom in the linear chain (from Sievers [12]).

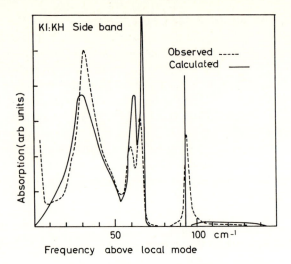

Fig. 2.4. Satellite lines associated with the U-center localized mode in KI measured by IR absorption (from McPherson and Timusk [40]).

vibrate in phase since we are in the acoustic mode region, whereas the impurity vibrates out of phase with them, yielding an oscillating electric dipole; thus it is possible to observe the mode in the acoustic region of the infrared. Also here, normally inactive modes in which the impurity is fixed may appear. Concluding, the linear model predicts the existence of localized, gap and resonant modes, associated with point defects in solids and particularly in ionic solids. These modes may be observed by IR absorption, or by other techniques which we shall discuss later.

2.2.2. The problem in three dimensions

The problem of resonant or localized excitations in a periodic structure is very general, since the excitations may be electrons in a metal (the point defect having a different number of electrons), a magnetic excitation associated with an impurity in a ferro- or antiferro-magnetic material, impurity localized excitons in an insulator etc. We shall not treat this vast subject in any detail and the interested reader may consult the literature previously cited [8, 10]. Formally the problems to solve for a vibrational excitation are:

given a point defect which perturbs the lattice, how to treat this perturbation in order to obtain the vibrational spectrum of the perturbed crystal;

knowing the perturbed crystal vibrational spectrum, how to calculate the variations in the various physical properties of the material: optical absorption, Raman effect, specific heat, thermal conductivity, spin–lattice relaxation etc.

A formalism using Green function methods has been developed for this kind of problem. Knowing the perfect crystal Green function, the perturbed crystal Green function may be calculated using some model for the elastic interaction between the defect and the lattice. Using this perturbed Green function, it is then possible to determine the influence of the impurity of the various properties of the crystal.

In the following sections of this chapter, we discuss typical examples of localized, gap and resonant modes: the OH^- system for localized modes, $KI:Cl^-$ and the F-center for gap modes, and finally the alkali-halides doped with Ag^+, Li^+ and OH^- for resonant modes.

2.3. Some examples of localized modes

The two examples we have chosen, the H^- and OH^- substitutional impurities, are very dissimilar. For the first, there are no new degrees of freedom and the localized mode corresponds to a vibration of the H^- relatively to the near ions in the lattice. For the OH^- impurity, there is the extra degree of freedom due to the intramolecular mode, the lattice acting only as a perturbation.

2.3.1. Hydrogen localized modes [15–18]

In alkali-halides, the substitutional H-ion is located at a site of cubic symmetry: for the usual alkali-earth fluorides, the symmetry is tetrahedral. For the interstitial H^-, the reverse is true.

2.3.1.1. The observation of localized modes associated with hydrogen in the alkali-halides and the alkali-earth fluorides. (a) The alkali-earth fluorides. Hydrogen and deuterium may be introduced in these materials by heating in the gas between 750 and 1000°C, with the sample in contact with metallic aluminium [19]. After treatment, the crystal does not show any EPR signal which could be due to neutral H. EPR measurements show that irradiation with X-rays at room temperature produces neutral interstitial atoms [19]. If the irradiation takes place at 77 K the atoms will be substitutional [20], and heating at 135 K will place them in the interstitial position. Before irradiation then a H^- substitutes a F^- [21]. We shall now analyse in detail CaF_2, although similar measurements have been performed for other hosts.

Hydrogen or deuterium in CaF_2 yield strong absorption bands in the

Table 2.1
Position and width (in cm^{-1}) of localized modes at 20 K associated
with substitutional H$^-$ and D$^-$, in CaF$_2$ [21].

Impurity	Line	Position	Width
H$^-$	Fundamental	965.5 $\pm$ 0.5	0.7[a]
	2nd harmonic	1919.8 $\pm$ 0.3	0.9
	3rd harmonic	2912.2 $\pm$ 0.5	3.0
		2825.6 $\pm$ 0.8	1.3
D$^-$	Fundamental	694.3 $\pm$ 0.5	2.2
	2nd harmonic	1384.5 $\pm$ 0.4	3.6
	3rd harmonic	2093.0 $\pm$ 1	5.4
		2047.0 $\pm$ 1	7.0

[a]Limited by the spectral resolution of the instrument.

near IR: a very important fundamental, the second harmonic and two third-harmonics (see table 2.1). The fundamental frequency ratio for H$^-$ and D$^-$ is equal to 1.39 instead of the value $2^{1/2}$ which would obtain if the oscillator was a perfect Einstein oscillator. This slight difference is due to anharmonicities which as we shall see broaden the absorption lines by allowing the decay of the excited vibrational system by phonon emission, thus decreasing the excited state lifetime. Anharmonicity is also responsible for the higher harmonics and for the temperature variation of peak position and band-shape for all these bands.

The neutral interstitial produced by irradiation (O$_h$ symmetry) shows only one (and less intense) absorption band, centered at 640 cm^{-1} with a width of 2.5 cm^{-1} at 100 K [22]. Although it is normal for the substitutional H$^-$ ion to have an IR active vibrational mode, it is less evident that the vibrations of the interstitial atom will couple directly with the electromagnetic field. However the atom is not a point, but rather a proton surrounded by an electronic cloud (cf. the shell model); such a cloud may move relative to the proton and this would create a dipole moment which would couple to the electromagnetic radiation. This effect is well known since even a rare gas atom impurity in a rare gas crystal will yield IR active modes [23]. For the case of H in CaF$_2$ the corresponding effective charge is 0.07e [22].

(b) The alkali-halides. Substitutional H (or D) in the alkali-halides shows a very different IR spectrum than that observed in CaF$_2$. At room temperature, only a broad featureless band is observed; when the temperature is lowered, a very narrow and intense peak appears in the middle of the band, accompanied by a very broad absorption region on the high energy side of the band, the intensity of which is about three orders of magnitude smaller than that of the narrow peak. Results on the "heavy" U-center give information on the origin of this spectrum: in KCl:D and KCl:H at low temperatures the peak frequency ratio are ω_L (H$^-$)/ω_L(D$^-$) = 1.40, ref. [7], more-

over, one of the satellite absorption bands is located 60 cm⁻¹ from the central peak in both cases. We may conclude then that the central peak is due to the excitation of the localized mode expected for a light impurity and that the satellite band is due to two quantum processes (simultaneous excitation of the localized mode plus a crystal phonon), allowed by anharmonic interactions Satellite absorption bands are also observed on the low energy side of the main peak. Their intensities decrease very quickly with temperature. These bands are equivalent to anti-Stokes Raman lines.

As for CaF_2, the anharmonicity of the potential for H^- causes several effects; however some of them are quite different in the case of the alkali-halides. In particular no second and third harmonics are observed. In both cases the localized mode shifts and broadens with increasing temperature: in fig. 2.5 we have reproduced the results of Dotsch et al. [24] for LiF and NaF; in the case of $NaF:H^-$ the width of the central peak behaves differently from other systems.

A priori, the attractive forces between H^- and the nearby ions should be practically the same as those between the halogen ion and the other ions, since they are much more dominated by the electrostatic interaction, and the impurity basically introduces only a mass defect. The frequency of the

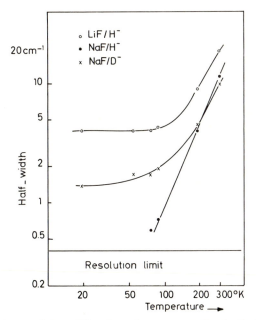

Fig. 2.5. Temperature variation of the absorption corresponding to the U-center localized mode in several alkali-halides (from Dötsch et al. [24]).

Table 2.2
Frequencies $\omega(H^-)$ expressed in cm^{-1} of the localized mode related to substitutional H^- in alkali-halides and ratio of $\omega(H^-)/\omega(D^-)$, $\omega(D^-)$ being the frequency of the localized mode related to substitutional D^- [24, 25].

Salt	$T(K)$	$\omega(H^-)$	$\dfrac{\omega(H^-)}{\omega(D^-)}$
LiF	20	1027	1.386
NaF	20	859.5	1.397
NaCl	90	563	1.38
NaBr	90	498	1.38
NaI	10	426.8	1.34
KCl	90	502	1.40
KI	4	378	–
RbCl	90	476	1.40
CsCl	100	357	–

localized mode $\omega(H^-)$ (see table 2.2) should be correlated with the frequency of the halogen ion in the perfect crystal at the edge of the Brillouin zone, where the alkali ions do not move:

$$\omega(H^-)/\omega(X^-) \sim (M_x/M_p)^{1/2}, \tag{2.21}$$

M_x being the halogen mass and M_p the proton mass. For NaI we have:

$$\omega(H^-)/\omega(X^-) \sim 5.7,$$

to be compared with

$$(M_x/M_p)^{1/2} \sim 15.1.$$

The error on ω_x cannot be larger than 40%; the disagreement indicates then that there is a force constant change when a hydrogen substitutes a halogen; we shall see that there is a reduction of the order of 0.3–0.4.

The interstitial H^- (U_1-center) also yields a localized mode; in KI Bauerle and Fritz [26] have observed only one mode at 718 cm^{-1} (width 0.8 cm^{-1} at 8 K) with a two-quantum sideband. Although this defect has T_d symmetry, no second or third harmonic absorption has been observed. The mode is difficult to observe because the U_1-center is produced by UV irradiation; in general, the H^- ion ejected into the interstitial position remains very near the vacancy it leaves behind: the ensuing perturbation lifts the degeneracy of the localized mode. To obtain a non-perturbed mode, the crystal must be heated slowly until all perturbed interstitials are eliminated.

2.3.1.2. Detailed study of hydrogen localized modes. (a) The frequency.
The first calculations of the localized mode frequency for the U-center in alkali-halides have been performed using a particularly simple model of

cubic diatomic lattices with nearest neighbor interactions only [27, 28].
Reasonable agreement with experiment has been obtained by assuming the
impurity to be a mass defect only. Unfortunately this model is not satisfactory for an ionic solid, and the frequencies calculated on the basis of the
rigid ion or shell models are about twice the experimental ones [29, 30].
Fieschi et al. [30] have calculated the localized mode frequency in lattice
models more compatible with ionic solids. The main conclusion was that a
large change in interactions took place upon substitution of a halogen ion
with H^-; it is thus impossible to consider the U-centers as mass defects.
For the rigid ion model, it is sufficient to introduce a nearest neighbor force-
constant change, which is due to change in the repulsive potential when H^-
replaces the halogen ion. This change is very difficult to determine theoretically since the interionic repulsive interactions are not well known. For
the shell model, there is of course a modification of the repulsive potential,
but also a variation in the electronic polarizability and of the effective charge
of the shell, and this makes the theoretical estimation even more difficult.
As we shall see these interaction changes are used as free parameters: in
particular, the ions nearby the U-center will have a vibration spectrum
which will be very different from that which they would have in the
perfect crystal: there will be gap and resonant modes; these are in turn
correctly interpreted using the force-constant changes determined from the
analysis of the localized mode [31].

(b) *The anharmonic effects.* In CaF_2, anharmonicity will cause higher
harmonics in the IR absorption spectrum of the impurity; in the alkali-
halides, sidebands of the localized mode appear. In the first case, it is
possible to interpret the results using an anharmonic oscillator in T_d symmetry; in the second case, we must consider all the crystal modes which can
couple to the localized mode.

The anharmonic oscillator and higher harmonics in absorption spectra.
Newman has treated the problem of the anharmonic oscillator in detail
[15] (see fig. 2.6), considering the anharmonicity as a perturbation of the
harmonic potential. In cubic symmetry the most general form for the anharmonic potential is (truncating at the 4th-order terms):

$$V = A(x^2 + y^2 + z^2) + C_1(x^4 + y^4 + z^4)$$
$$+ C_2(y^2z^2 + z^2x^2 + x^2y^2). \tag{2.22}$$

In tetrahedral symmetry this potential is:

$$V = A(x^2 + y^2 + z^2) + Bxyz + C_1(x^4 + y^4 + z^4)$$
$$+ C_2(y^2z^2 + z^2x^2 + x^2y^2). \tag{2.23}$$

The theory predicts several absorption bands corresponding to the sym-

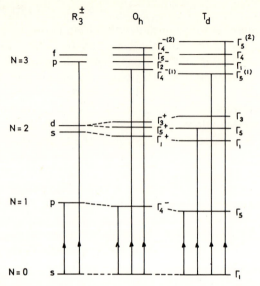

Fig. 2.6. Energy levels and allowed dipolar electric transitions for an anharmonic oscillator having the spherical ($R_3^{\pm}$), cubic (O_h) or tetrahedral symmetry (T_d) (from Newman [14]).

metry of the impurity: in cubic symmetry, a fundamental absorption and two third-harmonics, in tetrahedral symmetry: the fundamental, one second-harmonic and two third-harmonics.

For the U-center in alkali-halides (O_h symmetry) only the fundamental has been observed. This does not mean that there is no anharmonicity. More likely the oscillator strength for the transition is too small. In the preceding theoretical approach, we have not calculated oscillator strengths. On the other hand, the U-center in the alkali-earth fluorides shows all the transitions predicted by the theory; the B, C_1 and C_2 parameters can all be calculated from the three absorption bands. For CaF_2, Elliot et al. [21] have found:

$$B = 7.87 \times 10^{12} \, \text{erg cm}^{-2}$$

$$C_1 = -2.32 \times 10^{19} \, \text{erg cm}^{-2}$$

$$C_2 = -1.01 \times 10^{19} \, \text{erg cm}^{-2}.$$

In this particular case, the cubic term in the anharmonic potential is much greater than the quadratic terms because the first one splits the $N = 2$ levels by 48 cm^{-1}, where the terms C_1 and C_2 only give a contribution of 4.1 cm^{-1}.

Coupling with lattice phonons. The U-center is not considered as an anharmonic isolated oscillator of well defined symmetry, but rather as a system coupled to the crystal phonons by the anharmonic terms of the local potential. In O_h symmetry, the potential will contain both third- and fourth-order terms. The third-order terms consist of products $x_0^2 x_1$ and x_0^3 where x_0 is the displacement of the H^- ion and x_1 the displacement of the nearby ions. In the harmonic approximation, the lifetime of the localized mode is infinite since the defect may not return to the vibrational groundstate by phonon emission. In the anharmonic system, the lifetime becomes finite and the width of the line will increase as the lifetime decreases. At first this seems to explain the difference in the widths found for NaF:H and NaF:D (fig. 2.5). For NaF:H the energy of the localized mode (859 cm^{-1}) is much larger than phonon frequencies (< 300 cm^{-1}) so that an emission process involving two phonons is not allowed by energy conservation, whereas this process would be possible for the lower energy mode in NaF:D. For NaF:H it is necessary to have at least a three-phonon process to damp the localized excitation, and thus the width W will vary as $3\bar{n}^2 + 3\bar{n} + 1$ where $\bar{n}$ is the mean occupation number relative to the phonons involved in the damping; for a two-phonon emission event, W would vary as $2\bar{n} + 1$. The absence of two-phonon events explains the narrower width of the localized mode in NaF:H.

Unfortunately, this model does not correctly explain the variation of W with temperature: the curve (b) of fig. 2.5 shows the variation of $2\bar{n} + 1$ with temperature, which should describe the temperature variation of W for NaF:D. We must invoke a supplementary anharmonic mechanism to correctly interpret $W(T)$. This effect, proposed by Elliot et al [21] is well known in paramagnetic resonance: the Raman relaxation mechanism, i.e. the scattering of phonons by the localized excitation in our case. In this case, the excitation of the system consists of the localized mode and some crystal phonon: the inelastic diffusion of the phonon by the localized mode will yield a frequency fluctuation for the mode frequency, and thus a broadening. At high temperatures, this mechanism produces a temperature variation of W proportional to T^2, and this is effectively observed; at low temperatures, W should vary, as is the case in EPR, as T^7, and this is not at all supported by experiment. Thus we may conclude that at low temperature the width is mainly due to an emission of two or three phonons and that at high temperatures the phonon occupation number is sufficiently high for the scattering mechanism to become important. Several authors have interpreted all the $W(T)$ variations using these two mechanisms [21, 24, 32]. (cf. full line curve in fig. 2.5).

Anharmonic coupling and effect of uniaxial stress. The effect of uniaxial stress on the central peak is a direct measure of the crystal anharmonicity [24, 33, 34]. A general stress may be described by the appropriate linear combination of acoustic phonons of well defined symmetry and infinite wavelength. In O_h symmetry, these phonons will have the symmetries Γ_1^+ or

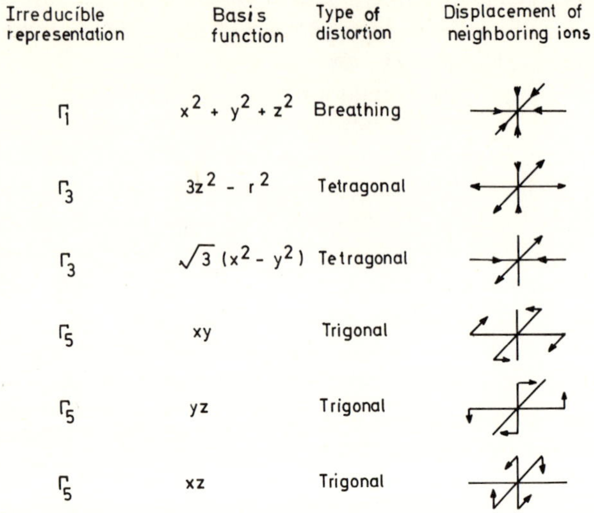

Irreducible representation	Basis function	Type of distortion	Displacement of neighboring ions
Γ_1	$x^2 + y^2 + z^2$	Breathing	
Γ_3	$3z^2 - r^2$	Tetragonal	
Γ_3	$\sqrt{3}\,(x^2 - y^2)$	Tetragonal	
Γ_5	xy	Trigonal	
Γ_5	yz	Trigonal	
Γ_5	xz	Trigonal	

Fig. 2.7. Representation of Γ_1^+, Γ_3^+ and Γ_5^+ normal modes in the O_h symmetry.

A_{1g} (breathing mode), Γ_3^+ or E_g (tetragonal mode) and Γ_5^+ or T_{2g} (trigonal mode); the corresponding deformations are shown in fig. 2.7. The coupling between the localized mode and these acoustic modes is described by third- and higher-order terms in the potential, so that the entire effect of the uniaxial stress will be connected with such terms.

A compression P_k causes a deformation field e_{ij}, in which the coefficients are given by Hooke's law:

$$e_{ij} = -\alpha_{ijkl}\, P_{kl}, \tag{2.24}$$

(here, we define a stress as a negative tension).

In eq. (2.24), the coefficients P_{kl} can be easily calculated for an arbitrary direction of the stress P:

$$
\begin{aligned}
P_{xx} &= \alpha^2 P & P_{yz} &= \beta\gamma P \\
P_{yy} &= \beta^2 P & P_{zx} &= \gamma\alpha P \\
P_{zz} &= \gamma^2 P & P_{xy} &= \alpha\beta P,
\end{aligned}
\tag{2.25}
$$

where α, β, γ are the direction cosines defining the orientation of P. In O_h symmetry, the elastic tensor may be written entirely in terms of three elastic constants σ_{11}, σ_{12}, σ_{44}.

One writes then the variation of the potential, $\Delta\Phi$, due to the deformation:

$$\Delta\Phi = \sum_{ij} b_{ij} e_{ij} = \sum_{\mu,v} b_\mu(\Gamma_v) e_\mu(\Gamma_v). \tag{2.26}$$

In eq. (2.26) we have introduced a tensor whose components transform as the μth basis function of the irreducible representation Γ_ν of the group O_h. The components of $e_\mu(\Gamma_\nu)$ are the following [25]:

$$
\begin{aligned}
e(\Gamma_1^+) &= e_{xx} + e_{yy} + e_{zz} = -(\sigma_{11} + 2\sigma_{12})(P_{xx} + P_{yy} + P_{zz}), \\
e_1(\Gamma_3^+) &= 2e_{zz} - e_{xx} - e_{yy} = -(\sigma_{11} - \sigma_{12})(2P_{zz} - P_{xx} - P_{yy}), \\
e_2(\Gamma_3^+) &= e_{xx} - e_{yy} = -(\sigma_{11} - \sigma_{12})(P_{xx} - P_{yy}), \\
e_1(\Gamma_5^+) &= e_{yz} = -\sigma_{44}P_{yz}, \\
e_2(\Gamma_5^+) &= e_{zx} = -\sigma_{44}P_{zx}, \\
e_3(\Gamma_5^+) &= e_{xy} = -\sigma_{44}P_{xy}.
\end{aligned}
\tag{2.27}
$$

The anharmonic parameters $b_\mu(\Gamma)$ are then linear combinations of the third derivatives of the potential Φ of the U-center. There is only one parameter for each irreducible representation, of the form [34]:

$$
\begin{aligned}
b(\Gamma_1^+) &= \frac{1}{6} \sum_i (\Phi_{xxx_i})x_i + 2(\Phi_{xxy_i})y_i, \\
b(\Gamma_3^+) &= \frac{1}{12} \sum_i (\Phi_{xxx_i})x_i - (\Phi_{xxy_i})y_i, \\
b(\Gamma_5^+) &= 2 \sum_i (\Phi_{xyx_i})y_i,
\end{aligned}
\tag{2.28}
$$

where x, y, z, are the U-center displacements along the cubic axis, x_i, y_i and z_i the position of a generic ion in the cubic lattice, the U-center being the origin. When there is only a nearest neighbor interaction, the eq. (2.28) becomes:

$$
\begin{aligned}
b(\Gamma_1^+) &= \tfrac{1}{3} r_0(\alpha_1' + 2\gamma_1'), \\
b(\Gamma_3^+) &= \tfrac{1}{6} r_0(\alpha_1' - \gamma_1'), \\
b(\Gamma_5^+) &= 4r_0\beta_1';
\end{aligned}
\tag{2.29}
$$

with $\alpha_1' = \Phi_{xxx_1}$, $\beta_1' = \Phi_{xyx_1}$ and $\gamma_1' = \Phi_{xxy_1}$.

Using eqs. (2.26) and (2.27), we can predict the absorption shifts for a given uniaxial stress and light polarization. The results are shown in table 2.3.

These predictions are actually verified: the stress lifts the degeneracy of the $N = 1$ state of the localized mode. With light propagating along one of the axes perpendicular to the stress direction, there are generally two bands, polarized parallel and perpendicular to the stress, as can be seen in fig. 2.8. The band shift is proportional to the stress intensity, within experimental error. Using these measurements, Fritz et al. [34] have calculated the $b(\Gamma_\nu)$ coefficients (table 2.4) of the U-center in KCl. In the same table, we also

Table 2.3
Theoretical displacement of the U-center localized mode, under a uniaxial stress.

Orientation of the stress	Polarization of measuring light	$-\dfrac{\Delta\omega}{P} \times 2m\omega_2$
[100]	[100]	$2b(\Gamma_1^+)(\sigma_{11} + 2\sigma_{12}) + 8b(\Gamma_3^+)(\sigma_{11} - \sigma_{12})$
	[010]	$2b(\Gamma_1^+)(\sigma_{11} + 2\sigma_{12}) - 4b(\Gamma_3^+)(\sigma_{11} - \sigma_{12})$
[110]	[110]	$2b(\Gamma_1^+)(\sigma_{11} + 2\sigma_{12}) + 2b(\Gamma_3^+)(\sigma_{11} - \sigma_{12})$ $+ \tfrac{1}{2}b(\Gamma_5^+)\sigma_{44}$
	[1$\bar{1}$0]	$2b(\Gamma_1^+)(\sigma_{14} + 2\sigma_{12}) + 2b(\Gamma_3^+)(\sigma_{11} - \sigma_{12})$ $- \tfrac{1}{2}b(\Gamma_5^+)\sigma_{44}$

show the results of Hayes and McDonald for the U-center in CaF$_2$[33].

We have discussed this type of experiment in some detail since it is the first example of perturbation spectroscopy, and particularly of piezospectroscopy, in this book. It is important that the reader follows the various analytical steps in calculating the deformation e_{ij} or $e_\mu(\Gamma_\nu)$ as a function of the applied stress. Nevertheless, this type of measurement only gives limited information. The results of table 2.4, in principle would yield the anharmonic coefficients α_1', β_1', γ_1' [eq. (2.29)], which could be compared with those obtained by other methods (for instance the B coefficient in the case

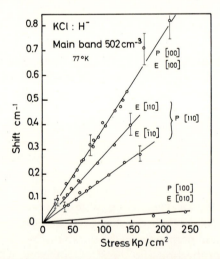

Fig. 2.8. Displacement of the U-center localized mode under an uniaxial stress along [100] or [110] measured with light polarized parallel or perpendicular to the stress (from Fritz et al. [34]).

Table 2.4
Parameters $b(\Gamma_\nu)$ of U-center obtained from piezospectroscopic measurements [33, 34].

KCl (O_h)	CaF$_2$ (T_d)
$-b(\Gamma_1^+) = 2.6 \times 10^4 \,\text{erg cm}^{-2}$	$-b(\Gamma_1) = 5.2 \times 10^4 \,\text{erg cm}^{-2}$
$-b(\Gamma_3^+) = 0.8 \times 10^4 \,\text{erg cm}^{-2}$	$-b(\Gamma_3) = 0.6 \times 10^4 \,\text{erg cm}^{-2}$
$-b(\Gamma_5^+) = 0.35 \times 10^4 \,\text{erg cm}^{-2}$	$-b(\Gamma_5) = 6.3 \times 10^4 \,\text{erg cm}^{-2}$

of CaF$_2$:H, on p. 52). Unfortunately, we have made a gross approximation which makes such comparison meaningless: using eq. (2.24), we have assumed that the crystal deformation about the U-center is the same as in the pure crystal, and this is certainly false since there is a relevant force-constant change around the U-center. This limitation can be found in all piezospectroscopic measurements, and thus it is useful to be cautious in the interpretation of the results. Nevertheless, we can draw a basic conclusion from table 2.4, namely that in O_h symmetry, it is the totally symmetric mode (Γ_1^+) which contributes most to the anharmonicity and the shear mode has little influence; this is not the case for T_d symmetry. This conclusion is physically satisfying.

Anharmonic coupling and satellite bands. The presence of satellite bands is one of the most direct evidences for anharmonicity. The events giving rise to it, involving one phonon and two vibrational excitations, are only subject to energy conservation, since the presence of the impurity destroys the crystal translational symmetry and thus relaxes the wavevector conservation rule. Thus phonons of all wavevectors may in principle contribute to the satellite bands, if the local crystal deformation corresponding to the two excitations yields a resultant dipole moment which can couple to the electromagnetic radiation; thus there will still be symmetry selection rules for the event. Such selection rules for satellite bands are simple; only even modes can contribute. The problem is formally the same as that of an electric dipole transition for a system coupled to lattice phonons. Since the localized-mode energy is much larger than that of lattice modes, we may use an adiabatic approximation to separate the impurity motion from that of its neighbors; the transition matrix elements will then be:

$$\langle i | g | \mathbf{r} | u \rangle | f \rangle, \tag{2.30}$$

where $| g \rangle$ and $| u \rangle$ describe the initial and final localized state (even and odd respectively), and $| i \rangle$ and $| f \rangle$ represent the crystal modes; at low temperatures $| i \rangle$ is even (zero-point energy). Since $\mathbf{r}$ is odd, the integral (2.30) is nonzero only if $| f \rangle$ is even; thus only even modes may contribute to the satellite lines. Finally, several calculations [36–40] have shown that the satellite bands are due to anharmonic coupling between the localized mode and its near neighbors; thus the even modes which can contribute to the sidebands will be those connected with motions near the impurity. In order to interpret the experimental results it is necessary to determine the vibrations in the neighborhood of the U-center, perturbed by the U-center itself.

Considering the perturbed motion of the nearest neighbors, Timusk and Klein [36] have reproduced to some extent the band shape of the first peak in the acoustic region of the satellite bands. This agreement was obtained without any adjustable parameters, the only parameter in the calculation being the force constant between H^- and the nearest neighbors, the value of which is obtained from the localized mode frequency. However, some predictions of their calculation are in disagreement with some experimental findings; in particular for KBr, resonant modes in the optical phonon region and a gap mode are predicted, and no such modes are observed experimentally. This model also does not predict some peaks which do appear on top of the acoustic bands. Actually, this model is too simple and its crude approximation for the coupling of H to its neighbors yields large errors in treating the high frequency vibrations involving the neighbors. The model has thus been improved by various authors by describing the local interaction more correctly. Among the several works on the subject, we shall discuss two briefly. Gethins et al. [39] have considered the lattice distorsion around H; this considerably improved agreement with experiment. In particular the satellite bands have been interpreted qualitatively for KBr and KI. A further improvement was added by McPherson and Timusk [40], who used a shell model for the H ion; the calculation gives good results by assuming that the shell is rigid but can move relative to the core of the defect, thus using only one adjustable parameter (fig. 2.4). The results and calculations on the satellite bands due to anharmonicity show that the vibration spectrum of the crystal near the impurity is different from the same bands for the pure crystal, which is to be expected. Since the coupling between H and its neighbors is much less than the pre-existing one in the pure crystal, and the masses of the ions are unchanged, these latter ones will vibrate at lower frequencies. We also understand the appearance of satellite bands for the localized or the resonant modes, and also the activation of odd parity modes in the IR [41]; for instance, the U-center in KI has IR resonant modes at 60 cm^{-1}.

(c) Perturbation of localized modes by impurities. The U-center localized mode is triply degenerate; a uniaxial stress may lift such degeneracy, but it can also be lifted by a nearby impurity. Several works have appeared on this point [41], and we shall discuss only one of them: that on H^-H^- (or D^-D^-) pairs in the alkali-halides. De Souza et al. [42] have succeeded in producing these pairs in the following way: at room temperature the U-centers are transformed into F-centers by X or UV irradiation. In this process, the

Table 2.5
Frequencies (in cm^{-1}) of IR active localized modes of H$^-$H$^-$ and D$^-$D$^-$ pairs in KCl [42].

U-center (H$^-$)	Pair H$^-$H$^-$	Mode	Pair D$^-$D$^-$	U-center (D$^-$)
	463.5	L	331.5	
502	535	T$_1$	375.5	358
	512.5	T$_2$	368	

hydrogen leaves the anionic site as a neutral atom and rapidly recombines with another hydrogen to form a stable interstitial molecule [43]. These molecules are mobile in the lattice at room temperature; they cannot re-combine with F-centers for lack of room, but may instead recombine with F$_2$-centers to form H$^-$H$^-$ pairs in the $\langle 110 \rangle$ direction. If the F$_2$-centers had been oriented previously, then the pairs also will be oriented in the same direction. The H$^-$H$^-$ pair has six non-degenerate normal modes since it has D$_{2h}$ symmetry. The out-of-phase modes will not be IR active since the dipole moments induced by the motion of the two H ions will mutually cancel. The three in-phase modes will be IR active, one being polarized along the defect axis L-mode, the other two perpendicularly to it (T$_1$ and T$_2$-modes). For randomly oriented pairs, there are three localized modes in KCl (table 2.5); using polarized IR light and oriented pairs, de Souza et al. have determined the symmetry of these modes. These results are very satisfying, since the localized mode degeneracy is lifted so that the frequency diminishes for a vibration in the pair direction, and increases for the two perpendicular directions. Unfortunately, no Raman measurements are available for the IR inactive modes.

2.3.2. The OH$^-$ localized modes in the alkali-halides

The OH$^-$ ion has a permanent dipole moment, which makes this defect para-electric; moreover it deforms the crystal anisotropically, and thus it is also para-elastic [44, 45]. The free molecular ion has three optical excitations: electronic, vibrational and rotational. In the crystalline matrix, it seems that the electronic and vibrational excitations are modified only slightly, whereas the rotational excitations are strongly perturbed by the crystal field. In most alkali-halides, the vibrational excitation is located at about 3600 cm^{-1} and the rotational excitation between 300 and 430 cm^{-1}. these frequencies are higher than the host-crystal vibrational frequencies, and thus these excitations yield localized modes. Nevertheless these modes are different from those of the U-center discussed previously, since they originate from the internal degrees of freedom of a molecule, and not from

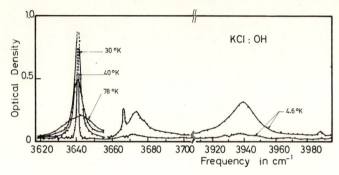

Fig. 2.9. IR absorption of OH⁻ doped KCl. The lowest curve corresponds to a concentration of 250 ppm and the highest to 1500 ppm (from Wedding and Klein [46]).

external modes. This is reflected in the theoretical approach and in this system the crystal lattice will appear only as a perturbation; this was certainly not the case for the U-center.

The vibrational absorption [44, 46†] consisted mainly of a band in the near IR due to the vibrational elongation of the molecule (stretching mode). At low temperature this band is so narrow that its width has not been measured yet. On its high energy side there is a satellite band at about 300 cm⁻¹ from it (fig. 2.9). The position of the stretching mode changes by a factor of $2^{1/2}$ upon substitution of OH⁻ with OD⁻, which indicates that the oxygen is practically fixed relative to the vibrating H ion (table 2.6). Also the energy difference between the principal line and the sideband changes by about $2^{1/2}$ upon isotropic substitution. Thus the sideband corresponds to the excitation of the stretching mode plus a libration mode of the molecular ion.

Table 2.6
Stretching and librational modes of the OH⁻ substitutional in several alkali-halides at low temperature [46, 48].

Crystal	Stretching mode		Libration mode	
	Position (cm⁻¹)	Oscillator strength $f \times 10^3$	Position (cm⁻¹)	Width (cm⁻¹)
NaCl	3654.5	0.65	396	18
KCl	3641	4.7	300	20
KCl(OD⁻)	2684.5	2.5	–	–
KBr	3617.5	3.3	313	6.1
KBr(OD⁻)	2668.0	1.7	–	–
KI	3603	3.0	283	4.4
RbCl	3632.5	2.5	272	13
CsBr	–	–	289	1.7

† An excellent bibliography on OH⁻ alkali-halides can be found in this reference.

The stretching mode frequency varies little from crystal to crystal (table 2.6); it diminishes regularly and slowly as the crystal lattice parameter increases. This vibrational mode is then practically unaffected by the surrounding crystal. The oscillator strength of the transition is particularly weak; for $\langle 100 \rangle$ oriented dipoles, which is the case here, the oscillator strength f, for a vibration–rotation transition ($J = 0 \rightarrow J = 1$) is given by: [46]

$$f = \tfrac{1}{3}(1/e)(dP/dR)^2, \tag{2.31}$$

where P is the molecular dipole moment relatively to the center of mass, and R the internuclear distance. This formula correctly interprets the oscillator strengths.

The stretching mode band has a width which increases with increasing temperature. The thermal processes responsible for this broadening must be multiple since there is no simple analytic expression to describe the temperature variation of the width. It is certain that the system has a resonant mode, i.e. it may interact with the crystal phonons. In the sideband in KCl there is a band at about 30 cm^{-1} from the central peak; this band is found as a direct absorption in the far IR. The low lying energy levels corresponding to this transition have also been studied by thermal conductivity measurements, since the lattice phonons of energy close to 30 cm^{-1} are strongly scattered. Thus the lifetime of the excited vibrational state is limited by the resonant scattering of these phonons; this Raman process has already been discussed for the U-center.

At very low temperatures the lifetime becomes sufficiently long for the probability of radiative excitation not to be ridiculously small relative to that for vibrational de-excitation. Capelletti et al. [47] have tried to observe this emission by pumping the system through its electronic transition in the visible, they claim to have observed a very weak emission. This preliminary result is encouraging since it indicates that weakly coupled systems may be studied not only in absorption but also in emission (very long vibrational lifetimes have also been observed in KCl:NO_2 by Rebane et al. see chapter 4).

The high energy sideband corresponds to a two-mode excitation, namely the stretching and the libration modes. The direct excitation of this latter mode has been observed in the IR [48, 49]. This system has no inversion symmetry center and all modes are Raman active as well as infrared active. Then it is normal to find in the far infrared modes observed in the stretching sideband; this was not the case for the U-center. Contrary to the stretching mode, the energy of the libration mode varies strongly with the host crystal (table 2.6); this shows that the crystal field influences this mode strongly. This absorption band always exists together with a sideband displaced by about 100 cm^{-1}; the sideband has a structure which varies from crystal to crystal. In order to interpret these results we shall use two different models. In the first the OH^- ion is considered as a quasi-free rotor whose motion is not perturbed by the crystal field. This model was developed some time ago by Devonshire [50] who calculated the effect of an octahedral field on the energy levels of a free rotor. This model has been reconsidered by Sauer [51] and applied with success to CN^- and NO_2^- in the alkali-halides [52–55], and the CH_4 molecule in the rare gas solids [56]. However this model cannot correctly interpret the properties of the substitutional OH^- [48, 49] in particular, it does not predict correctly either the position of the absorption peaks, or the observed isotropic displacements; moreover, it predicts absorption bands corresponding to various higher harmonics which have never been observed. So the system must be described as follows: the crystalline potential is great relative to the thermal energy of the molecule which will then librate slightly around its equilibrium position: the problem may then be reduced to the harmonic oscillator in one dimension (or two if account is taken of the two librational degrees of freedom of the molecule). Without entering into details, we may conclude that the molecule spends most of the time in a well defined equilibrium position. It is however well known that OH^- may pass from one equilibrium position to the other by tunnelling [44]. Application of a uniaxial stress or of an electric field lowers the local symmetry and the molecules will tend to align in the field direction (para-elastic and para-electric defects). These measurements yield the direction of the transition dipole moment; for the stretching mode, the transition takes place only if the electric field vector of the electromagnetic field is oriented along the molecular axis; thus this measure determines the orientation of the molecule in the crystal. It is also possible to determine the tunnelling jump frequency which varies from 10^{-2} to 10^9s^{-1} in the various alkali-halides studied. These results indicate that the Devonshire model of the perturbed rotor is not reasonable if the molecule oscillates more then 10^3 times about its equilibrium position before tunnelling to another equilibrium position. These results also indicate that the simple model of an oscillator about a single equilibrium position is not correct either. Tunnelling yields an energy separation of the levels, which is the greater the higher the jump frequency is. Thus it is reasonable that the width varies with the tunnelling frequency in the different alkali-halides [48].

In conclusion, alkali-halides doped with OH^- form a system which is much more complicated than the U-center since there are the extra degrees of freedom of the internal molecular vibrations. For the U-center, one must calculate the force-constant change which will modify the vibrational spectrum of the ions near the H and yield resonant modes. For OH^-, the two localized modes may be correctly interpreted using a very simple model in which the molecule has stretching and libration modes. It is however evident that there will also be a force-constant change for the OH^- molecular ion. Thus resonant modes should also exist for this system, and, as we shall see later on in this chapter, they do effectively exist.

2.4. Some examples of gap modes

It is a little academic to separate too rigidly the various modes associated with impurities in alkali-halides. In fact, some impurities feature simultaneously localized and resonant or gap modes or the same impurity in one of the alkali-halides will yield a localized mode and in another one a resonant or a gap mode. In fact, gap modes have properties which are very similar to resonant modes, with the exception that the width of the absorption bands is smaller since the lifetime of the excited state is longer. In table 2.7 we have reported the different experimental results for the U- and F-centers in crystals with a gap between acoustic and optical phonons. We see in particular that it suffices to add one electron to an F-center to change a gap mode into a resonant mode.

Before discussing in detail some examples of gap modes, we want to remark that color centers have been studied little in the IR, and for instance the IR absorption of the F-center has been studied only in KI and KBr; however some Raman scattering measurements are also available (see § 2.6.2).

As for localized modes, gap modes may appear in systems which do not introduce new degrees of freedom; for instance in KI, Cl on substitution of I yields two absorption bands, one in the gap at 77 cm^{-1} the other at 60 cm^{-1}, in the region of acoustic phonons [59]. High resolution measurements for the gap mode show that its absorption band is actually made up of two separate

Table 2.7
Gap and resonant modes associated with light impurities (hydrogen) or F-centers in several alkali-halides observed by IR absorption.

Crystal	Defect	Frequency (cm^{-1})	Mode
KBr [57]	F-center	99	Gap mode
KI [58]	F-center	83	Gap mode
KI [59]	F$^-$-center	61	Resonant mode
KBr [57]	U-center	89	Resonant mode
KI [58]	U-center	61	Resonant mode
KBr [57]	H$^-$-interstitial	98.7	Gap mode
KI [58]	H$^-$-interstitial	86.7	Gap mode

bands at 76.79 and 77 cm^{-1} [60]. Each band corresponds to an isotope of chlorine (^{35}Cl and ^{37}Cl). This isotopic effect had led to interesting theoretical results [61] since models which do not assume a force-constant change between the defect and its neighbors yield no isotope effect. In order to correctly interpret these experiments, it is necessary to take into account the lattice relaxation around the impurity and the resulting force-constant changes, not only for the nearest neighbors but also for the next nearest neighbors etc.

The F-center in KI [62] or KBr provides another example of a gap mode. In fig. 2.10, we see that the absorption band in the far IR (most far IR measurements are performed with a Fourier interferometer and Golay cell detectors. Resolution may be greatly improved by using a bolometer in either liquid ^{4}He, or better, ^{3}He [63]) is made up of three bands A, B, C which have the following positions: A, 82.62 cm^{-1}; B, 81.98 cm^{-1}; C, 81.19 cm^{-1}. The line at 80.05 cm^{-1}, is probably due to an F_A-center mode. The presence of three bands is due to the two stable ^{39}K and ^{41}K stable potassium isotopes in the crystal. Each of the lines may be attributed to a particular configuration in which zero, one or two of the six nearest neighbors of the F-center are replaced by the heavier isotope. This isotopic effect allows the determination of the force-constant change between the defect and the

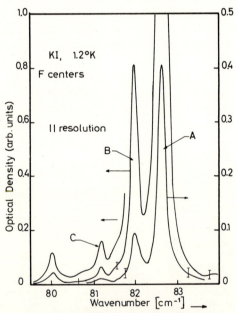

Fig. 2.10. Far infrared absorption of F-centers in KI for three different concentrations (from Bauerle and Hübner [62]).

most immediate shells of neighbors around it, and also between the shells themselves. Bauerle and Hubner [62] have used the Green function formalism. They introduced ionic relaxation about the defect, and force-constant changes between the vacancy and the first shell (A_{01}) and also between the first and fourth shells (A_{14}), using thus a model first considered by Benedek and Mulazzi [64] to interpret Raman scattering results; the difference is that they used a shell model, whereas Benedek and Mulazzi used the deformation dipole model. The isotropic effect may be interpreted if A_{01} and A_{14} are different from zero. From the calculation, the force-constant changes would be:

KI: $A_{01} = -0.50$ and $A_{14} = -0.060$

KBr: $A_{01} = -0.50$ and $A_{14} = +0.002$.

Thus in calculating the properties of defect modes it is convenient not to limit the defect space, since there may be force-constant changes induced by the defect extending over many lattice shells. This result is not unexpected since ENDOR measurements show the electron wavefunction to be still finite at the eighth shell (p); it is also well known that the F-center causes a strong ionic relaxation around itself, and this relaxation causes strong force-constant changes even at large distances from the defect [65, 66]. The importance of lattice relaxation has also been emphasized in calculations of other systems: KI:Cl and KI:Rb [68].

There may also exist gap modes associated with molecular impurities in substitutional positions; this is the case for CN^- and NO_2^- in KI [67].

2.5. Some resonant modes[†]

A great number of resonant modes are now known to exist in doped alkalihalides [9, 10, 16, 45, 63]. We shall discuss only a few examples: the Ag^+ (no new degree of freedom) and Li^+ (off-center ion) ions, and the OH^- molecular ion in substitutional position in the alkali-halides.

2.5.1. Ag^+ in the alkali-halides

Substitutional Ag^+ in the alkali-halides yields a very complicated far IR absorption spectrum (fig. 2.11). For KI:Ag [69], it consists of a very narrow line at low energy (17.3 cm^{-1}), two higher energy peaks at 55.8 and 63.5 cm^{-1} corresponding to maxima in the phonon density of states of the perfect crystal [70], and two asymmetric bands at 30 and 44.4 cm^{-1}

[†]See the review paper: A. S. Barker and A. J. Sievers, Rev. Mod. Phys. 47 (1975) 1.

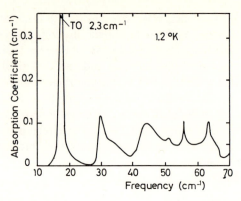

Fig. 2.11. Far infrared spectrum of KI:Ag$^+$ (from Kirby [69]).

which correspond to a two-quantum excitation as we shall see later. Uniaxial stress effects on the 17.3 cm^{-1} line show that it is due to a $A_{1g} \rightarrow T_{1u}$ transition in O_h symmetry. This line corresponds then to a quasi-localized mode in which the Ag$^+$ ion moves relative to its neighbors. From table 2.8, we see that the width of the resonant mode is much smaller in KI than in KBr: the phonon density of states is indeed weaker at the resonant mode frequency in KI than in KBr; thus the lifetime of the vibrational excited state will be longer for KI than for KBr.

The resonant mode line in KI is sufficiently narrow to give the possibility of measuring the isotope shift in passing from ^{107}Ag to ^{109}Ag [75]. The frequency ratio is:

$$\omega(107)/\omega(109) = 1.008 \pm 0.002$$

which is very close to the square root of the mass ratio, 1.009. This result confirms that this resonant mode is indeed connected with Ag$^+$, and that this system may be considered as an Einstein oscillator.

Uniaxial stress effects [71] in KI:Ag$^+$ indicate that the resonant mode is more strongly coupled with tetragonal lattice distorsion, and progressively

Table 2.8
Position ω and width at half-maximum of the absorption band related to the Ag$^+$ substitutional resonant mode in some alkali-halides.

System	ω(cm^{-1})	$\Delta\omega$(cm^{-1})	Ref.
NaCl:Ag$^+$	53	0.22	[72, 73]
KBr:Ag$^+$	33.5	0.25	[69, 74]
KI:Ag$^+$	17.5	0.05	[69, 74]

less with totally symmetric and trigonal vibrations respectively. The predominance of coupling to torsional modes indicates the considerable role played by non-central forces in this system; these forces must originate from some covalency in the bonding of Ag^+ with its neighbors (repulsion and ionic forces are central). A theoretical treatment of this system will be satisfactory only if non-central forces are added to the usual central force-constant changes. If uniaxial stress experiments allows the most complete determination of the local symmetry, the range of stress is severely limited by crystal strength to about 0.1% change in lattice parameter. This change is too small to appreciably alter the local potential. Hydrostatic pressure permits the existence of uniform strains and large changes of the lattice parameter (up to 2%). Kahan et al. [77] have performed a detailed study of the hydrostatic pressure effects on far IR properties of lattice resonant modes. The strain can be related to the pressure by an equation of the form:

$$dV/V = 3(S_{11} + 2S_{12})P - 3\delta P^2, \tag{2.32}$$

where the first pressure coefficient is the isothermal compressibility and second non-linear term is typically 5% of the first term for pressure around 5 kbar. Then, it is possible to measure the variation of the resonant mode absorption for different strains. Fig. 2.12 shows from the shift of the resonant mode of $KI:Ag^+$ the non-linear dependence of the resonant mode force-constant that Kahan et al. have obtained. Without discussing this result in detail, it is quite clear that the potential of Ag^+ in KI is not very well

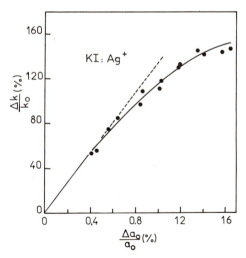

Fig. 2.12. Non-linear dependence of the resonant mode force constant in $KI:Ag^+$ with hydrostatic strain (from Kahan et al. [77]).

harmonic and that it is possible to extract from such experiments its anharmonic contribution having the A_{1g} symmetry.

Up to now, we have not discussed the temperature variation of gap-mode bands; measurements of this type are virtually impossible due to the strong increase with temperature of the fundamental crystal absorption. The situation is different for low frequency resonant modes. For KI: Ag, and likewise in practically all systems which feature a resonant mode, temperature increase has the effect of displacing the absorption band towards lower energies, of broadening it, and finally of decreasing the oscillator strength of this transition so that the band is not detectable any longer above 40 K. These effects are practically the same as those observed for zero-phonon lines in electronic transitions (cf. p. 113), which are in turn identical with Mössbauer lines [78]: the variation in oscillator strength is due to the modulation of the energy of the resonant mode by the lattice modes. As we shall show in chapter 3, if a Debye spectrum is assumed, which is reasonable at these low frequencies, the integrated intensity $I(T)$ of the absorption band should vary according to the well known equation:

$$I(T) = \exp\{a[1 + b(T^2/\theta_e^2)]\}. \tag{2.33}$$

The a and b coefficients are defined in the articles of Alexander et al. [76], θ_e is the effective Debye temperature. This equation does not completely account for the observed $I(T)$. For more quantitative agreement with experiment it is necessary to introduce a linear coupling between the odd resonant mode (T_{1u}) and an even resonant mode through the anharmonic terms in the potential [76]. For KI: Ag$^+$, thermal conductivity measurements have demonstrated the existence of a resonance at about 11 cm^{-1} [79].

Kirby [80] has detected the even resonant modes in the far IR for KI: Ag. An electric field should mix even and odd modes since the field is an odd parity perturbation; thus the even mode, containing some odd character, will absorb in the IR. Kirby has effectively observed two new absorption bands in the far IR at 15 and 25 cm^{-1}; these bands shift and their intensity increase as the applied field increases so that it is possible to extrapolate their position at zero field, i.e. the frequency of the pure even modes. Using symmetry considerations Kirby has shown that the mode E_g (Γ_3^+) has an energy of 16.35 cm^{-1} and the mode A_{1g} (Γ_1^+) an energy of 25 cm^{-1}. Thus the T_{1u} and E_g modes are nearly degenerate, and this justifies to some extent the conclusions of Alexander et al. on the variations of the oscillator strength for the resonant-mode absorption band. Finally the bands at 30 and 44.4 cm^{-1} (fig. 2.13) correspond to the sum events $T_{1u} + E_g$ for the first and $T_{1u} + A_{1g}$ for the second. The even resonant modes had been observed pre-

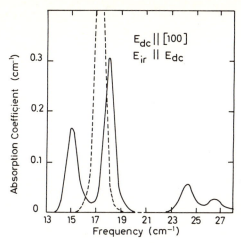

Fig. 2.13. Effect of an applied electric field to the FIR spectrum of KI:Ag$^+$ dotted line, zero field, full line, 130 kV cm^{-1} (from Kirby [69]).

viously only by thermal conductivity measurements at low temperatures [81] (their frequency is generally too low for Raman observation).

The thermal conductivity of a solid may be expressed as:

$$K = \frac{1}{3} \int_q \frac{dC}{d\omega} v^2 \tau(\omega) \, d\omega,$$ (2.34)

where C is the specific heat, v the phonon group velocity at frequency ω, and wavevector q, and τ the related relaxation time ($\Lambda = v\tau$) where Λ is the mean free path of the phonon). In fact, the relaxation time is a combination of partial relaxation times:

$$\frac{1}{\tau_{tot}} = \sum_i \frac{1}{\tau_i}.$$ (2.35)

To obtain the defect characteristic relaxation time, it is thus necessary to measure the thermal conductivity of the crystal and to determine the effects of surfaces and umklapp processes.

The defect associated relaxation time may be calculated in the harmonic approximation using Green functions. One obtains in general expressions of the type (see the review papers in ref. [81])

$$\tau^{-1} = A\omega_0^4 / [(\omega_0^2 - \omega^2)^2 + \Gamma\omega^6].$$ (2.36)

This equation describes an elastic resonant scattering event, with ω_0 being the resonance frequency. It is possible to determine A and Γ using experimental results and certain approximations (on force-constant changes, mass changes), i.e. obtain ω_0. As can be seen, this method is indirect; on the one hand it uses a macroscopic measurement in which all scattering events take part, on the other hand it necessitates particular hypotheses to determines the resonance frequency. The effect of Ag^+ on the thermal conductivity has been measured in several alkali-halides [82, 83]. In KI:Ag Bauman and Pohl [83] have calculated a frequency of $12 \, cm^{-1}$; considering the approximations used, the result is in excellent agreement with the optical determination of the mode E_g ($16.35 \, cm^{-1}$) of Kirby. In this system it seems than that the decrease in thermal conductivity is due essentially to the elastic scattering of the crystal phonons by the resonant level. This result is important because, in general, authors have a tendency to neglect the even resonant modes which are particularly difficult to observe, and then to launch themselves in complicated calculations to interpret thermal conductivity curves considering only the IR active odd modes.

2.5.2. Li^+ in alkali-halides: an off-center system

Substitutional Li^+ in alkali-halides should yield a high frequency localized mode since its mass is much smaller than that of the ion it replaces. In fact, this impurity causes a low frequency IR absorption corresponding to a resonant mode. For instance in KBr and NaCl one observes a low frequency absorption peak with weak bands at higher energy, with the peak width varying with Li concentration. In KCl, the situation is more complicated (fig. 2.14): the absorption band shifts to lower energies as Li^7 is replaced by Li^6, and it is composed of two bands; finally the absorption displays re-

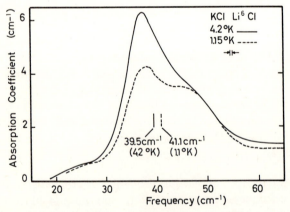

Fig. 2.14. FIR absorption of a KCl crystal doped with 2.4×10^{18} $^6Li^+$ ions (from Kirby et al. [84]).

Table 2.9
FIR absorption band at low temperature related to Li$^+$ substitutional in several alkali-halides [67, 84].

System	NaCl:Li^6	NaCl:Li^7	KBr:Li^6	KBr:Li^7
Position (cm^{-1})	45.3	44	17.71	16.07
Width (cm^{-1})	–	0.7	1	1

markable variations on lowering the temperature from 4.2 to 1.5 K (fig. 2.14).

We have already mentioned that it is necessary to describe a substitutional impurity both by a mass and a force-constant change. For Li$^+$, the mass decreases considerably; thus only a very substantial force-constant decrease can lead to such low frequencies as are observed. Sievers and Takeno [85] have shown that the force constant Li–Br may not be larger than 0.6% of the force constant K–Br of the perfect crystal. This result is reasonable in the sense that the ionic radius of Li$^+$ (0.78 Å) is much smaller than that of K$^+$ (1.33 Å). These simple considerations will lead to the assumption that Li$^+$ will be in a nearly square potential, i.e. the potential will be strongly anharmonic. In this hypothesis the system is not an Einstein oscillator any more and the isotopic shift of the absorption band is not given by $(7/6)^{1/2}$, as experimentally confirmed.

The system KCl:Li$^+$, studied in detail by Kirby et al. [84] shows many differences relative to NaCl:Li$^+$ and KBr:Li$^+$. Besides the ones described above, thermal conductivity measurements at very low temperatures [86] (between 0.05 and 4 K) show a discrepancy with respect to the ordinary T^3 behavior shown by the other systems: this anomaly may be described by the elastic diffusion of phonons on the very low frequency resonant mode, with the frequency being higher for Li^6 than for Li^7. This isotopic effect indicates that the resonant mode is indeed related to the Li$^+$ motion. The ensemble of the findings can be interpreted by assuming that Li$^+$ is not stable in the vacancy, and that it has eight equivalent off-center equilibrium positions in the $\langle 111 \rangle$ directions [45] and that it may tunnel among them. This model is well verified by many experiments (dielectric loss, adiabatic cooling) which will be discussed at the end of this chapter.

In the simple case of a double-well potential (fig. 2.15), the oscillator levels will be split in two, the separation increasing with the tunnelling probability. In this model, the mode observed at 40 cm^{-1} would correspond to the transition between the first two vibrational levels in the well, and the anomalies in the very low temperature thermal conductivity would be due to scattering of phonons of energy near that of the tunnelling splitting in the groundstate. Actually, the system is more complex since there are eight potential wells, each of them with C$_{3v}$ symmetry. Gomez et al. [87] have

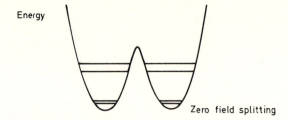

Fig. 2.15. Energy levels in a potential double well.

calculated the position of the tunnelling levels by assuming that Li^+ may tunnel only in the $\langle 100 \rangle$ direction; the groundstate would be composed of four sub-levels regularly spaced by energy Δ (zero-field separation). In order to completely interpret the IR absorption data, Kirby et al. [84] have calculated, in the framework of this model, the tunnelling induced splitting in the first excited state; they have also deduced selection rules for optical transitions among these sub-levels. In particular, there are thermal population effects between 4.2 and 1.5 K on the sub-levels of the groundstate ($\Delta \approx 0.6\,cm^{-1}$); their analysis yields a correct interpretation of the experimental results.

The off-center model has been verified by several methods. Using a paramagnetic phonon detector Walton [88] has determined Δ directly by measuring the frequency of the phonons diffused on the various sub-levels. An applied electric field in a specific crystallographic direction will lift the degeneracy of the eight potential wells: e.g. if the field is applied in the $\langle 100 \rangle$ direction, there will be four equivalent wells at higher energy than the remaining four. It is then possible to calculate the level separation as a function of applied field (fig. 2.16); group theory giving the symmetry of each level. Allowed electric dipole transition between some of these levels may be observed in the far IR or in a resonant cavity (para-electric resonance). Kirby et al. [84] have in fact, observed an absorption band in the far IR which shifts linearly to higher energies as the field intensity is increased (fig. 2.17).

Para-electric resonance is a more powerful technique since its sensitivity is better than IR absorption. Its principle is simple: in a EPR cavity which resonates at a specific frequency, the magnetic field is replaced by an electric field. This field splits the tunnelling levels of the groundstate and there will be a resonance signal whenever the energy separation of two opposite parity levels is matched by the phonon energy inside the cavity. Using several frequencies, it is then possible to build up the diagram of fig. 2.17. This very sensitive method has demonstrated that the model of Gomez et al. had to be improved by adding a finite tunnelling probability

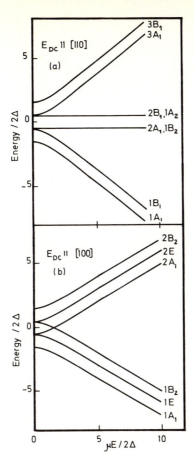

Fig. 2.16. Electric field effect on the multiplet structure of a tunnel system having eight potential wells in directions ⟨111⟩: (a) electric field along [110]; (b) electric field along [100]. (from Kirby et al. [84]).

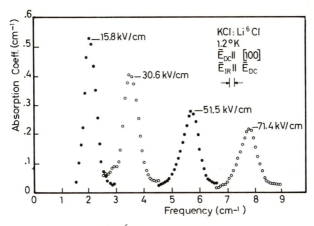

Fig. 2.17. FIR tunnel line in KCl:Li^6 for several values of an electric field applied along [100] (from Kirby et al. [84]).

in the $\langle 100 \rangle$ direction [89][†]. In the system KBr:Li$^+$, none of these preceding effects has been found. Nevertheless the isotopic shift of the resonant mode cannot be explained with a single potential well, even if highly anharmonic [90]. In this case, the Li$^+$ is still off-center, but the energy barrier between equivalent positions is very weak, smaller than the zero-point energy of the oscillator. Finally, several calculations have been performed to interpret these results [91, 92]. These calculations predict that the height of the potential barrier in KCl:Li must be very sensitive to the lattice parameter: Quigley and Das [93] also predict that the Li$^+$ ion must become centered under application of hydrostatic pressure of at least 7 kbar. In their hydrostatic pressure experiments, Kahan et al. [77] observed that the absorption spectra of KCl:Li$^+$ in the far IR at strains greater than 0.6 ($\approx$ 3 kbar) is very similar to the spectrum of KBr:Li$^+$ at zero strain. From these experiments, analysing the isotope effect in detail, they concluded that for both systems (KCl:Li$^+$ above 3 kbar and KBr:Li$^+$ at zero pressure) the impurity ion is in an intermediate configuration between the off-center and on-center limit. As in the case of KI:Ag$^+$, they were also able to determine precisely the level diagram of KCl:Li$^+$ and the anharmonic of each potential well. This level diagram is represented in fig. 2.18‡.

2.5.3. OH$^-$ in alkali-halides: another off-center system

The most detailed experimental study has been performed in NaCl:OH [84]. In this system, as in KCl:Li, two absorption bands appear at very low frequency (between 5 and 25 cm^{-1}) the intensity of which varies very strongly as the temperature varies from 4.2 to 0.6 K. This absorption spectrum also varies considerably with OH$^-$ concentration. In order to interpret these results, it is necessary that the OH$^-$ molecule changes its orientation by tunnelling, and also that the center of mass be displaced along a $\langle 100 \rangle$ direction, i.e. the ion be in a potential of symmetry C$_{4v}$ (fig. 2.19). The center of mass is off-center and thus there will be oscillations around its central position, which must be added to the librations. As for the preceding system, the tunnel effect will split the groundstate in several levels and the C$_{4v}$ symmetry for the OH$^-$ will lift the degeneracy of the first librational excited state. Finally, as for KCl:Li, it is possible to detect para-electric resonance [93]. The great sensitivity of the groundstate to the electric

[†] For references to para-electric resonance in KCl:Li$^+$ see ref. [89].
‡ However, F. Bridges and R. J. Russel [Solid St. Commun. 21 (1977) 1011] have observed a para-electric signal in KBr:Li$^+$ and they claim that the non-observation of this para-electric behavior came from the high concentration of hydroxyl ions in crystals which are associated with Li$^+$ impurities.

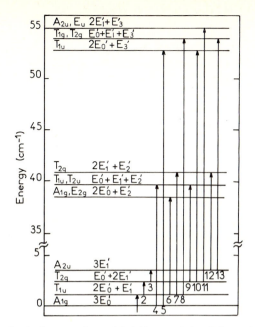

Fig. 2.18. Energy-level diagram of an eightfold potential well formed from a double-well potential $V(x)$ by the summation $V(x) + V(y) + V(z)$. The levels correspond approximatively to those of KCl: ^{6}Li^{+}. (from Kahan et al. [77]).

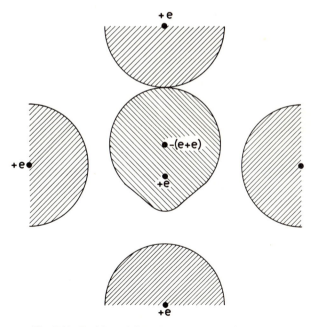

Fig. 2.19. Position of OH^{-} ion substitutional in KCl.

field explains the concentration effects; the polar molecule OH⁻ creates an electric field in the lattice, the average value of which is of the order of 200 kV cm⁻¹ for concentrations between 1500 and 2000 ppm. The model discussed here is still too simple to interpret all the experimental results; it appears that not only the center of mass is off-center in the ⟨100⟩ direction, but that it is necessary to add a small perpendicular displacement: this gives the center four equilibrium positions among which it will tunnel with a frequency much higher the "normal" tunnelling frequency among the different ⟨100⟩ orientations [94]. Raman scattering by OH⁻ in KCl and KBr [95] confirms these assumptions.

There are many other systems which have instabilities at low temperatures, and the interested reader will find a complete review in the article of Naranayamurti and Pohl [45].

2.6. Defect induced Raman scattering in ionic solids

2.6.1. Theoretical preliminaries

Since its discovery, the Raman effect has been extremely useful in the study of the structure and dynamics of matter. This type of measurement has found new life in the last fifteen years with the development of lasers and of sophisticated photon counting techniques. This new interest in Raman scattering has been particularly useful in the study of ionic solids, even in the alkali-halides, where Raman scattering is weak.

We may consider the Raman effect as an inelastic scattering event in which a photon of energy E_0 and wavevector k_0 is destroyed and a scattered photon of energy E_s and wavevector k_8 is created. The Raman effect is thus a two-photon process which must respect energy conservation:

$$E_0 - E_s = \pm E. \tag{2.37}$$

In a crystal, it must also satisfy the wavevector conservation law:

$$\mathbf{k}_0 - \mathbf{k}_s = \pm \mathbf{k}. \tag{2.38}$$

In eqs. (2.37) and (2.38), E is the energy and $\mathbf{k}$ the wavevector of the elementary excitation of the crystal (in our case, a phonon) created ($+$) or destroyed ($-$). The wavevector conservation which is directly related to the crystal periodicity considerably limits the type of phonons which can take part in the scattering event. The wavevector of a visible photon is very small relative to most of the Brillouin zone ($k \sim 10^{-3} \pi/a$); thus only phonons at or very near the center of the zone can participate in the Raman scattering.

If more than one phonon participates in the scattering event, it is the resultant wavevector that must be near zero. Thus for first-order Raman scattering (one-phonon event), only the optical phonons at the zone center (Γ) will contribute (the acoustic phonons yield the Brillouin scattering). In a crystal for which every lattice point is a center of symmetry, the symmetry selection rules and eq. (2.38) will forbid first-order scattering. Thus in the alkali-halides only second-order Raman scattering will be detected; in this case however the information about the lattice vibrations, although more abundant than for first-order scattering, is also much more difficult to unravel. If, now, the translational and point symmetry of the crystal are broken by the introduction of an impurity, eq. (2.38) does not hold and in principle phonons from the whole Brillouin zone will contribute to the scattering. However if the impurity remains at a center of inversion site, then only even phonons will be allowed to participate in the scattering; for such modes the impurity itself does not move; thus the resulting continuous spectrum will not be too dependent upon the particular impurity. The spectrum ·will be related to the one-phonon density of states of the crystal, multiplied by an appropriate electron vibration coupling function which will depend on the amplitude of the ionic movements near the impurity. This density of states is called the projected density of states. In the absence of defect induced modes, the projected density of states should be similar to the crystal density of states; thus the defect induced Raman scattering will yield information about the vibrational spectrum of the perfect crystal. If defect induced modes are present, then Raman scattering may detect them. Before giving some concrete examples, we shall briefly discuss the fundamental theory of Raman scattering by defects in crystals, using the approach discussed by Harley et al. [96].

If the phonon energy is small relative to the electronic energies of the crystal, but large with respect to vibrational energies (non-resonant scattering), then the intensity of the scattered light, at frequency ω_s, linearly polarized, per unit solid angle and unit frequency, is given by:

$$I(\omega_s) = \frac{\omega_0^4}{2\pi e^3} \sum_{\alpha\beta\gamma\lambda} n_\alpha n_\beta i_{\alpha\gamma,\beta\lambda}(\omega) E_\gamma(\omega_0) E_\lambda^*(\omega_0), \tag{2.39}$$

where: ω_0 is the incident light frequency, $\omega = \omega_s - \omega_0$, $\mathbf{n}$ is the unit vector of the scattered light parallel to the electric field, $E(\omega_0)$ the Fourier transform of the incident electric field, and the Raman tensor $i_{\alpha\beta,\gamma\lambda}$ is given by:

$$i_{\alpha\beta,\gamma\lambda}(\omega) = \frac{1}{2\pi} \int_{-\infty}^{+\infty} d\tau \, e^{-\omega\tau} \langle P_{\alpha\gamma}^*(\mathbf{u}, \tau) P_{\beta\lambda}(\mathbf{u}) \rangle. \tag{2.40}$$

The Raman tensor is the space–time Fourier transform of the polarizability correlation function, since $P_{\beta\lambda}(\mathbf{u})$ is the static polarizability operator for the electronic groundstate in a given nuclear position configuration $\{\mathbf{u}\}$ and $P_{\alpha\gamma}(\mathbf{u}, \tau)$ represents the time variation of such an operator. The instantaneous dipole moment may be written as:

$$\mathbf{M} = \sum_n z_n \mathbf{u}_n + \sum_l z_e \mathbf{u}_c,$$

where z_n and z_e, are the ionic and electronic charges respectively. The polarizability tensor is given by:

$$P_{\alpha\gamma}(\mathbf{u}) = \sum_\mu [\langle 0|M_\alpha|\mu\rangle\langle\mu|M_\gamma|0\rangle + \text{CC}]/\hbar\omega_{\mu 0}, \tag{2.41}$$

where $|\mu\rangle$ represents the electronic states of the crystal and $\hbar\omega_{\mu 0}$ the energy of the virtual transition to the level μ. Expanding $P_{\alpha\gamma}(\mathbf{u})$ in normal coordinates q_f, we obtain:

$$P_{\alpha\gamma}(\mathbf{u}) = P_{\alpha\gamma,0} + \sum_f P_{\alpha\gamma,f} q_f + \dots . \tag{2.42}$$

The correlation function in eq. (2.40) is then given by:

$$\langle P_{\alpha\gamma}^*(\mathbf{u}, \tau) P_{\beta\lambda}(\mathbf{u})\rangle = P_{\alpha\gamma,0} P_{\beta\lambda,0} + \sum_{ff'} P_{\alpha\gamma,f} P_{\beta\lambda,f'} \langle q_f(\tau) q_{f'}\rangle + \dots ,$$

where the fact that $q_f(t) \equiv 0$ in the harmonic approximation that has been used. It is possible to calculate the function $\langle q_f(t) q_{f'}\rangle$ in this approximation, and obtain for the Raman spectrum:

$$i_{\alpha\gamma,\beta\lambda}(\omega) = \hbar\sum_f (2\omega_f)^{-1} P_{\alpha\gamma,f} P_{\beta\lambda,f} [n(\omega_f)\delta(\omega + \omega_f) +$$
$$+ n(-\omega_f)\delta(\omega - \omega_f)], \tag{2.43}$$

where $n(\omega) = [1 - \exp(-\hbar\omega/kT)]^{-1}$. The term in $n(\omega_f)$ represents the Stokes event in which the phonon is created, the term $n(-\omega_f)$ corresponding to the anti-Stokes event in which a phonon is destroyed. The Raman tensor will be zero if the product $p_{\alpha\gamma,f} P_{\beta\lambda,f}$ is zero, and this will depend on the coefficients of the series (2.42). These coefficients are strongly tied to the crystal symmetry. For instance in a crystal with O_h symmetry, $P_{\alpha\gamma,f}$ is different from zero only if the distorsion represented by f has the symmetry A_{1g}, E_g and T_{2g}. Since in alkali-halides the only optical phonons at Γ have the symmetry T_{1u}, $i_{\alpha\gamma,\beta\lambda}(\omega)$ will be identically zero and there will be no first-order Raman scattering, as we had anticipated earlier. The even modes of A_{1g}, E_g and T_{2g} symmetry will be actived by the introduction of a substitutional impurity in the alkali-halide lattice. Thus the projected density of states in this case will refer to crystal phonons of the aforementioned symmetries; taking polarized spectra it is then possible to separate these different symmetry contributions to the spectrum.

2.6.2. *Raman scattering from the Tl impurity in alkali-halides*

In fig. 2.20, we show the polarized Raman spectra due to Tl in KI [96]; in this figure are also reported the theoretical predictions for the various polarizations and the spectrum for pure KI. It is important to note that the projected density of states has been obtained by assuming that there was no force-constant change; thus the agreement with experiment would indicate that the Raman spectrum reflects the perfect crystal density of states, and that Tl does not cause either localized or quasi-localized modes. Similar results have been obtained also for KBr:Tl and KCl:Tl [96]. From these experiments Harley et al. have also shown that by measuring the total Raman intensity for the various polarizations it was possible to determine directly the coupling of the electrons with the crystal phonons of symmetry A_{1g}, E_g and T_{2g}. This method is thus complementary in this respect with the measurements of electron–long wavelength acoustic phonon coupling performed by studying the effect of uniaxial stress on the optical bands of an impurity (chapter 3) [97].

2.6.3. *Raman scattering in mixed crystals*

The mixed ionic crystals yield an interesting example of disorder induced first-order Raman scattering [98–100], in the same fashion as the semi-

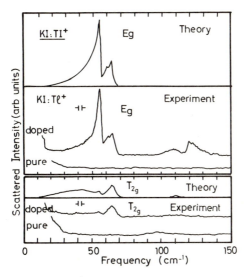

Fig. 2.20. Raman spectra of KI:Tl$^+$ measured at 15 K (a and b: E_g and T_{2g} symmetry contributions to the spectrum). (from Harley et al. [96]).

conductors $CdZn_{1-n}S_n$, ref. [101]. The mixed alkali-halides studied have been: KCl:Br, ref. [98], KBr:Cl, ref. [100] and NaCl:Br, NaCl:I, KCl:I and NaCl:Li, ref. [98]. The interest of these experiments is due to the fact that it is possible to obtain mixed crystals in all proportions, i.e. to go continuously for instance from KCl to KBr varying the Cl or the Br concentration. If the impurity does not induce localized or gap modes, the Raman spectrum of the crystal $AX_n Y_{1-n}$ will reflect the projected density of states of the crystal AX, when $(1 - n)/n = 1$ and that of the crystal AY when $n/(1 - n) = 1$. This is what is actually observed qualitatively [98, 99]. Actually the results are less clear than for Tl-doped alkali-halides, since it appears that resonant modes contribute significantly to the Raman spectrum. Another problem is caused by the non-negligible second-order contribution. The spectra for $KCl_n Br_{1-n}$ are represented in fig. 2.21 for several values of n; the spectra would approximate the density of states of KCl $(n \sim 0)$ and KBr $(n \sim 1)$.

The study of mixed crystals in which symmetry forbids first-order scattering in the ordered phase allows the study of the vibrational spectrum of the pure crystal or, at least, of the eventual defect modes which could not

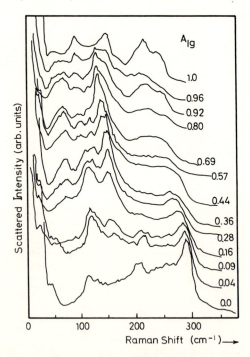

Fig. 2.21. A_{1g} contribution of Raman spectra in $KCl_n Br_{1-n}$ measured at room temperature. (from Nair and Walker [98]).

be seen in IR. It is on account of this for instance that Montgomery and Kirby [102] have detected two resonant modes of E_g and A_g symmetry in NaCl:Cu at temperatures between 4.3 and 394 K, which yield peaks at 36.6 cm^{-1} (E_g) and 46.5 cm^{-1} (A_g), which broaden considerably as the temperature increases. The existence of an even mode at about 30 cm^{-1} had been inferred from IR measurements [76].

The impurity induced Raman scattering has been detected in many other systems where the host crystal symmetry is not so high as for the alkali-halides. A great number of results concerns the II.VI semiconductors [103]; in this case, the measurements have revealed resonance effects which appear when the incident frequency approaches an electronic resonance of the system. We shall treat resonance Raman scattering more in detail in chapter 3.

2.7. Defect tunnelling in ionic solids[†]

In the preceding paragraphs we have shown that the far IR properties of KCl:OH$^-$ and KCl:Li$^+$ could be explained only by assuming a tunnelling rotation of the OH$^-$ ion or a tunnelling displacement of the Li$^+$ ion (in this last case the displacement being along one of the eight off-centers $\langle 111 \rangle$ directions around the equilibrium position). Besides inducing specific vibrational defect modes, these tunnelling defects feature a number of other interesting properties, which we shall discuss in this paragraph. In particular, these defects have in general lower symmetry than that of the host lattice, and therefore deform it anisotropically; an applied stress will orient them, i.e. they are para-elastic. Some of them also have a permanent electric dipole moment; they are then also para-electric and will orient in an electric field; thus they will modify the crystal dielectric constant.

2.7.1. Study of a para-elastic system; substitutional O_2^- in the alkali-halides

In a now classic work Känzig [104] has, using EPR, studied the properties of the molecular ion O_2^- substitutional to Cl$^-$ in several alkali-halides under applied stress (see table 2.10). In the free crystal the molecular ion, which is smaller than the ion it replaces (see fig. 2.22) is oriented along $\langle 110 \rangle$. The defect g tensor is anisotropic enough to allow the detection of the resonance

---

[†]For review papers on para-electric phenomena, mainly in alkali-halides see: V. Narayana-murti and R. O. Pohl, Rev. Mod. Phys. 42 (1970) 201; F. Bridges, Crit. Rev. in Solid St. Sci. 5 (1975) 1.

Table 2.10
The lifting of O_2^- orientational degeneracy by uniaxial stress. θ is the
angle between the stress direction and the molecular ion axis.

Stress direction		Energy levels	Degeneracy
[111]	35°, 26°	ΔU_1	3
	90°		3
[110]	0°	$\Delta U_3 \Delta U_2$	1
	60°		4
	90°		1
[100]	45°	ΔU_4	4
	90°		2

lines from ions oriented in different $\langle 110 \rangle$ directions. An external uniaxial
stress will lift the orientational degeneracy of the ions, and the resulting
level populations will be distributed according to Boltzmann statistics.

Having determined the different energy levels and their relative populations by EPR, the energies U may be calculated. It seems that the orientation
of the ions under stress is still $\langle 110 \rangle$ however, the energy differences U are
proportional to the applied pressure. Thus an elasticity tensor may be
defined which relates the defect energy to the surrounding lattice deformations:

$$U = b_1 \varepsilon_{\xi\xi} + b_2 \varepsilon_{\eta\eta} + b_3 \varepsilon_{\rho\rho} \tag{2.44}$$

where ε_{ij} is the relative crystal deformation along the principal axis j of the
defect. Assuming that the crystal elastic constants around the defect are the

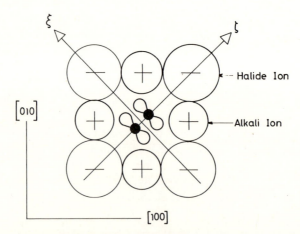

Fig. 2.22. The substitutional O_2^- ion in KCl, KBr and KI. The lobes of the p-wavefunction are
indicated schematically. (from Känzig [104]).

Table 2.11
b_j' values of the elastic tensor of O_2^- in several alkali-halides (unit: 10^{-12} erg).

	b_1'	b_2'	b_3'
KCl	−0.73	2.35	−1.62
KBr	−0.25	1.43	−1.18
KI	−0.01	1.27	−1.26

same as for the pure crystal, it is possible to calculate the b_1, b_2 and b_3 components of this tensor. Känzig also used another tensor which eliminates the isotropic part of the deformation:

$$b_j' = b_j - \tfrac{1}{3}(b_1 + b_2 + b_3) \tag{2.45}$$

and obtained the results reported in table 2.11.

The results in table 2.11 indicate that the O_2^- elastic energy is smallest for a stress along the η axis and largest for stress along the molecular axis ρ: thus the center will orient perpendicularly to the stress direction.

Using these methods, Känzig has studied the disorientation kinetics as the stress is lifted. He found that these kinetics (characteristic times inferior to about 50 s) depend on temperature, applied pressure, the type of host crystal etc. In particular:

the disorientation time τ (as the stress is lifted) is independent of the stress whereas the orientation time τ^* depends both on the magnitude and on the direction of the applied stress;

below 4 K, τ and τ^* increase proportionally to $1/T$, i.e. they behave as the T_1 relaxation time of a paramagnetic impurity.

From these results it is evident that the defect reorientation does not take place by thermal activation over a potential barrier: in this case in fact τ should be proportional to $\exp(W/kT)$, W being the barrier height. Thus the O_2^- ion jumps from one orientation to another by tunnelling, which must be phonon assisted since the rate depends on temperature. As in the case of spin relaxation, several possible mechanisms are possible: one-phonon or two-phonon (Raman) processes for instance. Only the direct one-phonon process yields a relaxation time which is inversely proportional to T [105–107]. In the case of orientation under stress (τ^*), Silsbee [107] has shown that the results could be interpreted by taking into account the energy difference between the initial and final states, i.e. that one- or multiple-phonon processes could be active. He has also shown that, in KI, the 60° rotation probability was essentially the same as the 90° rotation probability. Finally, later, Pfister and Känzig [108] have shown a large isotropic effect by comparing results for $^{16}O_2^-$ and $^{18}O_2^-$; this confirms

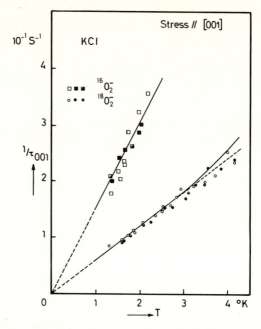

Fig. 2.23. Temperature variation of the O_2^- relaxation rate under zero stress for an initial stress in the [100] direction. In the figure an important isotropic effect of such a dependence is also shown. (after Pfister and Känzig [108]).

the tunnelling nature of the reorientation process. In fig. 2.23, such an effect is shown on the T dependence of the relaxation rate.

Another result of these works is the demonstration that it is incorrect to treat the system as a molecule rotating in a rigid potential; instead, as it turns the molecule distorts the surrounding lattice which in turn damps the rotation itself. This effect is similar to the polaron effect in which the electron polarizes the surrounding medium and is in turn slowed down by it. In fact a small polaron model which includes one-phonon relaxation channels yields good agreement with experiments. This system is also peculiar in the sense that the relaxation time is very long (about 10 s at 4 K compared to 10^{-8} s for $KCl:Li^+$ at the same temperature); thus the tunnelling splitting of the groundstate is of the order of 10^{-6} eV. This small splitting is difficult to detect since in any crystal there will exist residual internal strains of the order of 0.2 to 0.4 kg mm^{-2}, which will broaden the levels to a width of 10^{-4} eV.

Before concluding this paragraph, let mention a very interesting phenomenon observed by Silsbee and Bojko [109] for the reorientation of N_2^- substitutional in KI. The general behavior of this system is quite similar to the previous one with a quantum tunnelling very much faster than in KCl:

O_2^-. As in this last system, they observed the important consequences of the dynamic response of the host crystal to the motion of the impurity: a comparison of two different types of reorientation, one through 60° and one through 90° jumps measured by ESR, indicates that the tunnelling matrix elements are significantly altered by polarization of the host. The interesting phenomenon in this system is the following observation: in the helium temperature regime, the 90° reorientation demonstrates the influence of the symmetry of the nuclear spin states upon the allowed matrix elements for reorientation. It is not our purpose to explain their experiment in detail and we send the reader to their paper. The general idea of this phenomenon is the following: the ability of a homopolar diatomic molecule to change its orientation between two equivalent directions in the crystal may depend upon the relative orientation of nuclear spins of the two atoms of the molecule. This assertion, rather offensive to one's classical intuition, is a reflection of symmetry requirements on the tunnelling matrix elements connecting the two stable configurations, and is essentially a solid state analog of the familiar nuclear spin symmetry restrictions on the allowed rotational states of homopolar diatomic molecules (ortho and parahydrogen for example).

2.7.2. *The para-elastic and para-electric systems*

Para-electric systems such as $KCl:OH^-$ (permanent electric dipole moment) and $KCl:Li^+$ (dipole moment associated with the off-center position of the Li^+ ion) are generally also para-elastic since they have lower symmetry than the ions they substitute. These systems may be classified in off-center systems and rotational systems. In this subsection, we shall discuss only two prototypes for each class, i.e. substitutional Li^+ in KCl and substitutional OH^- in several alkali-halides.

2.7.2.1. *Stable positions of Li^+ in $KCl:Li^+$ and stable orientations OH^- in the alkali-halides.* In § 2.5.2 we have shown that the far IR absorption in these systems could be well interpreted by supposing that Li^+ jumped by tunnelling among eight equivalent $\langle 111 \rangle$ off-center positions and that the OH^- molecule (in a first approximation) rotated by tunnelling among six equilibrium $\langle 100 \rangle$ orientations (fig. 2.18). Since these systems are not paramagnetic, EPR could not be used to determine such equilibrium positions. For the case of Li^+, Sack and Byer [110] have determined the off-center direction by studying the ultrasonic attenuation of transverse waves propagating along the different crystallographic directions; using different polarization, they have shown that the elastic coefficient s_{44} varies with

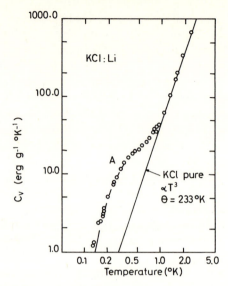

Fig. 2.24. The specific heat of KCl:Li between 0.14 and 1 K. The sample contained 2.8×10^{17} Li cm^{-3}. The full line represents the specific heat of pure KCl for $\theta_D = 233$ K. The dashed curve is calculated using a Schottky anomaly model (after Harrison et al. [111]).

Li$^+$ concentration, whereas the difference $s_{11} - s_{12}$ does not. Locally, a variation in s_{44} corresponds to a deformation of symmetry T_{2g} (fig. 2.7). Whereas a variation in $s_{11} - s_{12}$ would indicate a E_g deformation. Now a T_{2g} deformation is compatible only with an Li displacement along $\langle 111 \rangle$. It is interesting to mention the low temperature specific heat measurements [111]: in KCl:Li$^+$, there is a significant change in the specific heat relative to a pure crystal (fig. 2.24), between 0.15 and 1 K. The entropy change associated with a system of defects each of which has equal probability to be in any of the eight tunnel states is given by:

$$\Delta S = \int_0^\infty \frac{C_v}{T} \, dT = n_{Li} k \log 8. \tag{2.46}$$

The good agreement with the specific heat data indicates that all the Li$^+$ ions are displaced in one of the $\langle 111 \rangle$ directions [however the precision of the measurements is not sufficient to allow a determination of the number of equivalent sites (Lüty, private communication)].

These same methods may be used to determine the OH$^-$ orientation in the alkali-halides. It is however simpler to study the electric field or stress induced linear dichroism on the OH$^-$ stretching IR absorption [112]: by definition, the dipole moment associated to this vibration is in the direction of the molecule. By applying an electric field F parallel to the propagation direction of the IR incident light, Hartel [112] has observed, in RbCl and

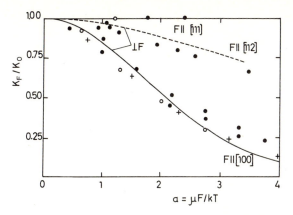

Fig. 2.25. Relative change of the OH⁻ stretching absorption in RbCl as a function of applied electric field and temperature for different orientations of the field intensity F (the field is parallel to the incident light propagation direction). The full and dashed curves represent the predictions of the model in which the dipoles are oriented along ⟨100⟩. ●, $T = 4.4$ K; +, $T = 3.2$ K; ○, $T = 2.7$ K; (from Hartell[112]).

KBr, a large absorption decrease for the field in the [100] direction, and no change for F in the [111] direction. This result clearly indicates that the dipoles are oriented along ⟨100⟩. The observed absorption changes in RbCl:OH⁻ are shown in fig. 2.25, versus $\mu F/kT$; the curves represent the best theoretical fit obtained assuming ⟨100⟩ dipoles reorienting in the electric field and obeying Boltzmann statistics [122]. The numerical value for the dipole moment used was $\mu = 0.6$ eÅ assuming that the local electric field was equal to the applied field. These measurements thus allow not only the direction of the molecular ion axis to be determined, but also enable a value for the ion dipole moment to be obtained. Similar results have also been obtained using an uniaxial stress perpendicular to the propagation direction of the incident light [112]. For [100] stress, the absorption reduces practically to zero for incident light polarized parallel to the stress, indicating an almost total alignment of the dipoles. The absorption instead increases by about 50% for perpendicularly polarized light (fig. 2.26). For [111] stress, there is no absorption change, and thus no reorientation. These results may be interpreted only for ⟨100⟩ oriented elastic dipoles, having the form of cigars aligned perpendicularly to the applied stress. Here also the results may be interpreted assuming an energy difference αS between dipoles aligned parallel to and perpendicularly to the applied stress and Boltzmann equilibrium between the two orientations. The results of fig. 2.26, yield α, which, for RbCl, is equal to 5.5×10^{-24} cm³.

2.7.2.2. Tunnelling effects. The two fundamental quantities describing a tunnelling system are the jump frequency (or relaxation time, which as we

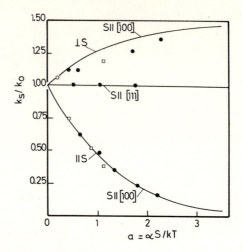

Fig. 2.26. Relative change of the OH⁻ stretching absorption in RbCl as a function of
applied stress and temperature for different stress directions and with light polarized parallel
or penpendicular to the stress. The full curves are calculated using a model of ⟨100⟩ oriented
elastic dipoles (OH⁻ concentration: 2×10^{-5}). ●, $T = 2.7$ K; □, $T = 4.4$ K; (from Hartel
[112]).

have shown can be as high as ten seconds in the case of $KCl:O_2^-$) and the
groundstate splitting, or more generally, its structure. The relaxation time
may be determined by several methods [45]. Here we shall limit ourselves
to the optical method used by Kapphan and Lüty [113] in OH⁻ doped alkali-
halides. The OH⁻ impurity is associated with an optical absorption band
in the UV whose oscillator strength is a tensor ($f_\sigma = 0.05$ and $f_\pi = 0.17$, i.e.
$f_\sigma/f_\pi = g = 0.3$; where σ and π are the parallel and perpendicular polarizations relative to the dipolar axis respectively). As in the case of the preceding
experiments, the application of an electric field in the direction of propagation of the incident light will decrease the absorption. Using Boltzmann
statistics to describe the level populations, the absorption $K(F)$ must obey
eq. (2.47):

$$\frac{K(F)}{K(0)} = \frac{3}{2(2+g)} \left[2 - \frac{4(1-g)}{e^a + e^{-a} + 4} \right], \qquad (2.47)$$

where $a = \mu F/kT$. This is effectively what is observed.

The relaxation time of the dipoles can be measured by following the
variations of $K(F)$ after a fast change in F. The relaxation time was found
to be strongly dependent on the host crystal (fig. 2.27); in any case, for T
below 5 K, it is proportional to T^{-1}, which indicates a prevalence of one-
phonon relaxation processes. Above 5 K, τ varies as T^{-4} (as determined for
the case of RbBr:OH⁻), indicating a two-phonon relaxation [113]. Finally,

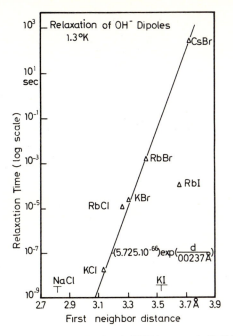

Fig. 2.27. The relaxation time of substitutional OH⁻ in several alkali-halides as a function of the nearest neighbor distance (from Kapphan and Lüty [113]).

these results also indicate that the reorientations proceed by 90° jumps only. In fig. 2.27, we see that, with the exception of the iodides, the relaxation time increases exponentially with the lattice parameter of the host crystal; this is reasonable since the potential barrier height will vary in proportion to the lattice parameter. In RbI and KI the results will be different since the OH⁻ dipoles are oriented along ⟨110⟩.

More detailed analysis of the tunnelling process of such systems can be found in the review paper by Bridges [114] as well as in a paper by Lüty [115].

2.7.2.3. Experimental determination of tunnel splitting. We have indicated several techniques to measure energy splittings of the multiplet of the fundamental state (quadruplet regularly spaced for KCl:Li⁺ in the O_h symmetry and a triplet for KCl:OH⁻ in the same symmetry): para-electric resonance (PER), far infrared absorption under electric field and scattering of phonons measured by thermal conductivity. Two other methods that are particularly interesting have been developed that we shall present now.

Phonon spectroscopy. A new kind of phonon spectrometer has been developed by Kinder [116] using a superconducting tunnel junction for generation and detection of phonons. In phonon generation, the junction is biased at some voltage V above the energy gap of the superconducting junction $2\Delta_G$. Electrons crossing the gap have kinetic energy which is quickly lost by phonon emission. The phonon spectrum is thus fairly broad with a sharp cut-off at a maximum energy $E_m = eV - 2\Delta_G$. Then, a small change in the applied voltage shifts the phonon edge, the rest of the spectrum remaining unchanged. By a small modulation of the applied DC voltage, a narrow band of phonons is turned off and on; a lock-in amplifying the signal given by the detector at the modulation frequency will analyze only modulated phonons at the energy E_m.

Windheim and Kinder [117] have measured directly the tunnel splitting in NaCl:OH$^-$ using this technique. In this experiment they used Sn–Oxide–Sn junctions as phonon generators emitting longitudinal phonons and Al–Oxide–Al as detectors and they observed the transitions between the A_{1g} and E_g states of the tunnel multiplets as it is shown on fig. 2.28, and its behavior with applied uniaxial stress. This method is very sensitive: from the observed absorption strength they have estimated the OH$^-$ concentration of their crystal to be of the order of 0.1 to 0.01 ppm.

Infrared spectroscopy with a microwave spectrograph. It is possible to directly obtain the zero-field splitting by measuring the absorption of microwaves as a function of frequency. Scott and Flygare [118] have performed this experiment for frequencies between 8 and 40 GHz on KCl:OH$^-$. They have observed several absorption bands main characteristics of which are reported in table 2.12.

If the OH$^-$ ions were located at a site of purely O_h symmetry, there would be only three levels, with the splitting between the ground level (A_{1g}) and the first excited level (T_{1u}) being twice that between the ground and second excited level (E_g) (fig. 2.28).

In this scheme, only the $A_{1g} \to T_{1u}$ and $T_{1u} \to E_g$ transitions are allowed. To account for the larger number of transitions actually observed, we must assume that the defect is not in cubic symmetry. Scott and Flygare have shown that a small C_{4v} distortion can account correctly for the microwave absorption: thus the OH$^-$ center of mass is off-center, as confirmed by the

Table 2.12
Microwave absorption peaks in KCl:OH$^-$ [118]

Frequency (GHz)	Relative intensity	Width (MHz)
8.35	0.5	150
9.5	5	150
11.5	10	200
14.2	1	100
22.3	2	150
22.8	0.5	100

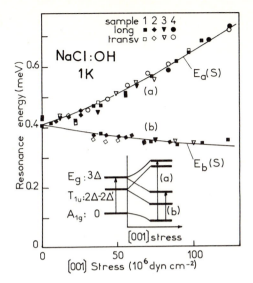

Fig. 2.28. Resonance energy as function of [001] stress for four different samples measured by phonon spectroscopy. Inset: level scheme in O_h symmetry and under stress (After Kinder [117]).

far IR absorption measurements (see §2.5.3). From table 2.12, we also observe that the widths of the bands are large: this is due to the internal stresses which are also known to broaden the zero-phonon lines in optical transitions (see chapter 3).

2.7.3. *Phenomena associated with para-electricity and para-elasticity*

Several effects well known in paramagnetic systems will also be found in para-electric and para-elastic systems. Thus we shall have adiabatic cooling by para-electric and para-elastic disorientation, dielectric constant variation due to para-electric defects and elastic constant variation due to para-elastic defects.

2.7.3.1. Adiabatic cooling [45, 114, 119]. Ordering and disordering of an ensemble of electric dipoles due to an applied electric field will yield an entropy variation, i.e. a temperature variation in adiabatic conditions. In the region where the total polarization P is given by $P = N\mu \cdot \mu E/kT$, temperature variations ΔT will be observed upon turning the electric field on and off which will be given by:

$$\Delta T = \pm N\mu^2 E^2 / 6 C_v kT, \tag{2.48}$$

the equation being valid only for $\Delta T \ll T$. This effect was first observed by

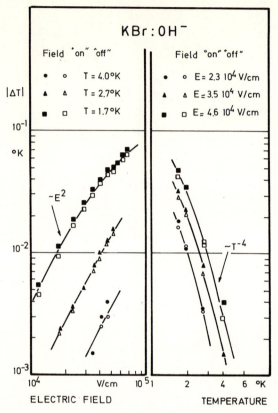

Fig. 2.29. Para-electric heating and cooling as a function of electric field and temperature (from Lüty [119]).

Sheperd and Feher [120] in KCl:OH⁻ under either electric field or uniaxial stress. These measurements were then continued by Lüty and collaborators [119] who showed that indeed ΔT is proportional to the square of the electric field and inversely proportional to the fourth power of temperature (let us recall that in this range $c_v \sim T^3$); furthermore the reversibility of the effects was also shown (fig. 2.29).

In the case of non-adiabatic experiments, where the time variation of the applied field is much shorter than the dipole relaxation time, there is no more reversibility: as the electric field is applied, all dipoles which are not already in the field direction will so orient and the energy thus given the crystal will be twice that given in the adiabatic process. As the field vanishes, all dipoles will be suddenly at the same energy: there will be no temperature variation and only the entropy will vary. Thus in the irreversible process the heating is twice that for the reversible process and the cooling is zero. One may also produce all the intermediate cases by varying the time dependence of the applied field using characteristic times τ_E about the relaxation time τ; in fig. 2.30, we observe the heating (higher curve) and cooling (lower curve) as a function of τ_E [119]. This behavior is in good agreement with the preceding predictions. From these experiments the values of τ were determined and found to be in good agreement with those determined optically. Finally, these experiments also yield μ [eq. (2.48)], by assuming that the local field is the same as the external field. Similar measurements have also been performed on other para-electric systems (see review papers previously mentioned [45, 114]).

2.7.3.2. The dielectric constant. In every case, the low temperature dielectric constant is found to increase. For a diluted system, (fig. 2.31a, a curve), the dielectric constant excess is well explained by a model of independent electric dipoles which orient along the field direction and disorient thermally [121, 122]: for this the Clausius–Mossotti equation and Lorentz local field are used:

$$\frac{\varepsilon - 1}{\varepsilon + 2} = \frac{\varepsilon_m - 1}{\varepsilon_m + 2} + \frac{4\pi}{3} N\alpha, \tag{2.49}$$

where $\alpha = \mu^2/3kT$ (classical Langevin equation), for $kT > \Delta$ (where Δ is the zero field splitting) and:

$$\alpha = (2\mu^2/3\Delta) \tanh (\Delta/2kT), \tag{2.50}$$

for $kT < \Delta$. In this calculation by Gomez et al. [123] the electric field is assumed to be weak. At high concentrations, the interactions between dipoles become important and collective effects must be considered.

2.7.4. Electric dipole interactions: local order at high concentrations

As the para-electric and para-elastic impurity concentration increases, there may be interactions between the defects which will modify some of the previously discussed effects. For para-electric impurities, there is a reduction of adiabatic cooling [124], a modification in the behavior of the dielectric constant (fig. 2.30), and a variation in the far IR absorption [125]. It is evident that these modifications are due to dipole–dipole interactions. The system is analogous to an ensemble of spins dispersed in a non-magnetic matrix. One of the fundamental problems in such systems is that of local order; this was discussed for instance by Anderson [126] for a magnetic system. Among the effects associated to such local order, we recall the residual magnetization which decreases with time, a specific heat propor-

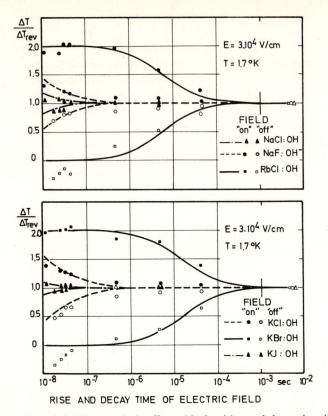

Fig. 2.30. Variation of the electro-caloric effect with the rising and decreasing time of the electric field for several OH⁻ doped alkali-halides (from Lüty [119]).

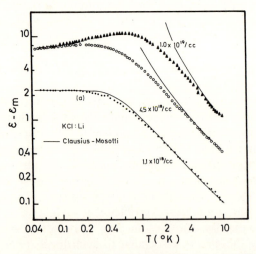

Fig. 2.31. Variation with temperature of $\varepsilon - \varepsilon_m$ in KCl:Li and pure KCl. Full lines are calculated from the Clausius–Mosotti equation using 0.82 cm^{-1} as the zero-field splitting of the groundstate.

tional to T and a susceptibility maximum at a temperature which varies linearly with concentration. Similar effects have been observed by Fiory [122] in some para-electric systems which we shall presently discuss.

In a KCl:Li which had been under electric field at very low temperature ($T \sim 0.2$ K) there remains a residual polarization which decreases slowly and linearly with log t (several minutes). This residual polarization is proportional to the square of the dipole concentration. The results indicate that the polarization decay is composed of several characteristic times, due to impurity pairs at different mutual distance. This interpretation is confirmed by the behavior of the ac dielectric constant as a function of the frequency v of the applied electric field: ε decreases about logarithmically with v at very low temperature. Finally, as for dilute magnetic systems, ε has a maximum at a temperature T_m (fig. 2.31) proportional to the impurity concentration

$$T_m = Np^2/k\varepsilon_m, \tag{2.51}$$

where p is the impurity dipole moment connected with the local moment μ by $p = \frac{1}{3}(\varepsilon_m + 2)$ (Lorentz correction). The energy Np^2/ε_m is just the dipole interaction energy for a pair separated by the average distance $r = N^{-1/3}$. This result indicates that the anomalies are indeed due to a pair interaction: when $T > T_m$, kT is larger than E_p and we have normal behavior, i.e. the pairs do not form stable configurations; for $T < T_m$, such configurations are instead stable. The dielectric constant decreases below T_m, indicating a possible antiferroelectric-type of ordering. The specific heat does not show any anomaly typical of a possible phase transformation; such an absence seems normal since the dipole interaction energy is distributed about a certain average and it is not constant. In the case of highly doped crystals, Fiory has observed an increase in c_v proportional to $T^{3/2}$, analogous to that due to spin waves in a ferromagnet. He then tried without success to find dipolar waves. These possible collective modes may be difficult to observe due to the strong damping caused by their large coupling to the lattice. Thus the existence of an antiferroelectric phase at very low temperature and very high concentrations is not demonstrated. This system is best viewed as an ensemble of statistically distributed dipoles with strong local ordering.

This ordering problem is a very interesting one because it is formally the same as the ordering of spin glasses, but with long range interactions whereas spin interactions are practically always short range (only magnetic dipolar interactions are long range). Recently, Fisher and Klein [127] have performed a theoretical analysis of the ordering process using the Mean Random Molecular Field Approximation. Within this model a new kind of phase transition should exist if the strength of the dipole field is larger than the tunnel splitting. In this phase, the polarization of the system vanishes, but short- and mean-range order exist below a well defined transition temperature T_c. The electric susceptibility varies continuously with temperature but its derivative should show a discontinuity at T_c. In this paper, they claimed that such a discontinuity was not observed by Fiory because he used a too large electric field which perturbed the system and destroyed the transition. This problem has been analyzed in detail by Benedek for different systems [128].

The systems we have described are in any case very interesting and have been extensively studied in recent years. For the off-center systems, a general theory of their symmetry lowering is still to appear. However, an analogy with the Jahn–Teller effect is possible (in which the electronic degeneracy is substituted by the orientational degeneracy); it has been recently proposed by Bersukher [138].

2.8. Phonon generation and detection

2.8.1. Generation of monochromatic phonons by microwaves

The para-electric defects we have discussed, have tunnelling levels with splittings of the order of or below about $2\,cm^{-1}$ and strong coupling with lattice phonons; the latter feature is not possessed by the spin systems which have splittings in the same energy range. The strong phonon coupling makes the defects efficient phonon scatterers. Furthermore the defects may couple to the phonons by their dipole moment and this allows the production of quasi-monochromatic beams of phonons in a crystal containing para-electric impurities. The experiment was first done by Channon et al. [129] by exciting the tunnelling levels of KCl:Li in a resonant microwave cavity. The reader may find the details of this experiment in the work of Channon [130] and Hertler [131] including the relative theoretical analysis: here we shall just postulate that the basic coupling mechanism between the tunnelling states and the lattice is due to the Coulomb interaction between the defect dipole moment and the electric quadrupole moment created at the defect site by the nearest neighbor ions. Since the phonons are created by a process of resonant scattering their frequency distribution width will be equal to that of the incident microwave beam, if the latter one is much larger than the inverse lifetime of the excited tunnelling level; in the opposite case, the width of the emitted phonon spectrum will be determined by the lifetime of the excited tunnelling level [132]. Such a lifetime is very short in KCl:Li ($\sim 10^{-8}\,s$) and this allows the production of a large number of phonons since saturation occurs only at very high powers; thus some milliwatts of monochromatic phonons were obtained by this technique [130].

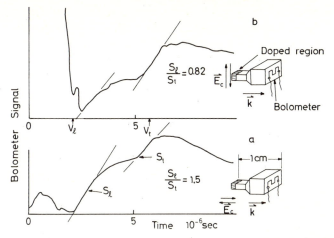

Fig. 2.32. Signal of the bolometer receiving phonons emitted by KCl:Li (left hand of the sample), the phonon wavevector k, being parallel to [100]. (a), k parallel to the cavity electric field; (b), k perpendicular to the cavity electric field. S_L and S_t are slopes corresponding respectively to longitudinal and transverse phonons (from Channon et al. [129]).

By using short microwave pulses, Channon et al. were also able to produce short pulses of ballistic phonons, by which they studied the creation process by observing the direction and the modes of the created phonons. In fig. 2.32, we have reproduced one of the figures from the work of Channon et al. which shows the signal given by the phonon detector as a function of time, after the crystal had received a microwave pulse at $t = 0$. The experimental device is shown in the same figure: after the crystal had been cut as shown, the lithium was introduced at one of the extremities by diffusion, and the phonon detector (a superconducting film of indium) was evaporated at the other extremity. The doped portion of the crystal was placed in a K-band microwave cavity, whereas the remaining part was outside; the ensemble was immersed in pumped liquid helium. The first linear rise in the signal corresponds to the arrival of the longitudinal phonons and the second rise to the arrival of the transverse ones; it is thus possible to determine sound velocities to frequencies up to some tens of GHz. Channon et al. have shown that the phonons are indeed emitted by the substitutional lithium by observing the effect of an applied electric field: the number of created phonons diminishes as the field increases, indicating that the field detunes the resonant levels from the microwave frequency.

These methods allow the production of monochromatic (about 30 GHz) phonon beams inside the crystal.

In particular one of the most serious difficulties for hypersound injection, related to the junction between the piezoelectric generator and the crystals, is eliminated. However to our knowledge this method has only been tested quantitatively in KCl:Li. In principle, all tunnelling systems could be used yielding a wide range of resonant frequencies.

So far we have only discussed phonon generation by para-electric impurities. It is well known that monochromatic phonons may be also generated by saturating a spin system in a magnetic field [133, 134]; however in this case the spin coupling, both to microwaves and to phonons is several orders of magnitude smaller than in the preceding case, i.e. the number of phonons created is much smaller. Recently, measurements have been performed using the principle of para-electric resonance [135]. A KCl crystal about 10 cm long is doped with lithium at both ends, which are then placed in a microwave cavity under an applied electric field. A microwave pulse in the first cavity generates monochromatic phonons which are then detected by para-electric resonance in the second cavity.

2.8.2. Phonon detection by optical methods

In the experiment described in fig. 2.31, the generated phonons were detected with a superconducting bolometer. In this paragraph, we shall describe optical phonon detection techniques which, being sensitive to phonon frequency, allow the performance of phonon spectroscopy. This feature is very important since apart from using the phonon spectrometer or superconducting tunnelling junctions to detect tunnel splitting, we have no method of directly determining the spectral distribution of phonons created in some way. The first phonon spectrometers were used by Anderson and Sabiski [135] and by Brya et al. [133]. These first authors have used as the phonon detector the Zeeman sublevels of the groundstate of the paramagnetic ion Tm^{+2} in SrF_2, whereas the latter authors have detected the phonons generated by the spin system in MgO:Ni using the corresponding Brillouin scattering. Here we shall discuss in more detail the method of Anderson and Sabiski which is more related to the subject matter of this book. They generated monochromatic phonons by saturating the Zeeman levels of the Tm^{+2} at one extremity of a cylindrical crystal, and then detected these phonons by measuring the paramagnetic part of the magnetic circular dichroism of the absorption band at 5800 Å associated with these ions [135]. We shall show in chapter 4 for the F-center that the paramagnetic magnetic circular dichroism is proportional to the spin polarization in the groundstate:

$$\langle S_t \rangle = A \tanh (g\beta H/2kT), \tag{2.52}$$

where $A = -\frac{1}{2}$ for the F-center case. This formula, valid in thermal equilibrium can also be applied to Tm^{+2}. As a monochromatic phonon beam arrives at the impurity site, the effective spin temperature of the spin system will change if the resonance condition $\omega_0 \sim g\beta H$ is satisfied. This was of course the case in the Anderson–Sabiski experiment, since the Tm ions act both as generators and detectors of phonons. It is easy to saturate the populations of the Zeeman sublevels even without using a resonant cavity since the relaxation time is very long (about 0.15 s at 30 kG and 1.3 K); this long relaxation time also makes more simple the detection of a change in S, the magnetic circular dichroism signal. Anderson and Sabiski have shown that the spin temperature variation δT associated with a variation δS is given by the formula:

$$\delta T = -2(H/T \cdot \partial S/\partial H)^{-1}\delta S, \tag{2.53}$$

where H is the magnetic field and the factor two is due to the fact that of two resonant lines only one is saturated [135]. In fig. 2.33, we have shown the results of the spin temperature variation as a function of distance between the generator and the detector. The phonon beam intensity decreases exponentially; such a decrease is found to obey the law:

$$K(\partial^2\rho/\partial x^2) - \tau^{-1}\rho + g = 0, \tag{2.54}$$

where ρ is the density of non-equilibrium phonons, x the distance, g the

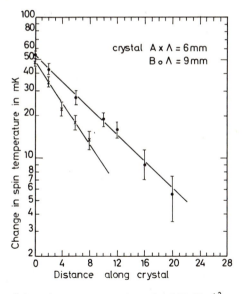

Fig. 2.33. Variation of the spin temperature along the SrF_2:Tm^{+2} crystal, phonons being injected at 1.3 K according to the method described in the text (from Anderson and Sabisky [135]).

phonon generation rate and τ their lifetime. It is possible to calculate ρ from the variation of the effective temperature and g (at saturation) from the impurity concentration and the relaxation time T_1 so that it is possible to obtain τ for phonons of frequency ω and spectral width $\delta\omega$. In one of their samples of $SrF_2:Tm^{+2}$, Anderson and Sabiski have determined the following values: $\tau \approx 12.5 \ \mu s$ and $\Lambda = 9$ mm (Λ: mean free path). This value of Λ indicates that phonon scattering in this system is due essentially to reflection at the walls.

The methods described here could be extended in several ways. For instance, by coupling the generator and detector the spectral distribution of phonons could be determined since the Zeeman splitting may be varied from zero in zero field to some cm^{-1} for a field of some tens of kG. The phonons could be generated by heat pulses and their spectra then determined by such a method. The DCM techniques may be then very useful in what we shall call very low frequency spectroscopy. Compared to other phonon spectroscopy techniques [137, 138] this one has the advantage of being extremely sensitive, since the actual measurement involves spectroscopy with visible light. It is however evident that this limits it to transparent solids.

References to chapter 2

[1] M. Born and K. Huang, Dynamical Theory of crystal lattices (Oxford, 1954); A. A. Maradudin, E. W. Montroll, G. H. Weiss and I. P. Ipatova, Theory of lattice dynamics in the harmonic approximation (Academic Press, New York, 1971, 2nd ed.).

[2] A. D. B. Woods, B. N. Brockhouse, R. A. Cowley and W. Cochran, Phys. Rev. **131** (1963) 1025.

[3] W. Cochran, Phonons in perfect lattices and in lattices with point imperfections (Oliver and Boyd, 1966) p. 62.

[4] E. W. Kellerman, Phil. Trans. **A238** (1940) 513.

[5] A. M. Karo and J. R. Hardy, Phys. Rev. **129** (1963); Phys. Rev. **141**A (1966) 696.

[6] J. Dick and W. Overhauser, Phys. Rev. **112** (1958) 90.

[7] G. Schafer, J. Phys. Chem. Solids **12** (1960) 233.

[8] A. A. Maradudin, Rep. Prog. Phys. **28** (1965) 331; Solid St. Phys. **18** (1966) 273; I. Lifshitz and A. M. Kozevitch, Rep. Prog. Phys. **29** (1966) 217; M. V. Klein, in: Physics of Color Centers, (op. cit.) p. 429.

[9] Phonons and phonon interactions, ed. T. A. Bak (Benjamin, New York, 1964); Elementary Excitations in Solids, ed. R. W. H. Stevenson (Oliver and Boyd, 1966); Phonons in perfect lattices and in lattices with point imperfections, ed. A. A. Maradudin and G. F. Nardelli (Plenum, New York, 1969).

[10] Localized Excitations in Solids, ed. R. F. Wallis (Plenum Press, New York, 1968).

[11] P. Mazer, E. W. Montroll and R. B. Potto, J. Wash. Acad. Sci. **46** (1956) 2.

[12] A. J. Sievers, in: Localized Excitations in Solids (op. cit).

[13] L. Genzel, H. F. Renk and R. Weber, Phys. Stat. Solidi **12** (1965) 639.

[14] R. Brout and W. M. Visscher, Phys. Rev. Letters **9** (1962) 54.

[15] R. C. Newman, Adv. Phys. **18** (1969) 545.

[16] M. V. Klein, Physics of Color centers (op. cit).

[17] A. A. Maradudin, Solid St. Phys. vol. 19 (1968).
[18] B. Fritz, Localized Excitations in Solids, ed. Wallis (Plenum Press, New York, 1968) p. 480.
[19] J. L. Hall and R. T. Schumacher, Phys. Rev. 127 (1962) 1892.
[20] R. G. Bessent, W. Hayes and J. W. Hodby, Phys. Letters 15 (1965) 115; Proc. R. Soc. A297 (1967) 396.
[21] R. J. Elliott, W. Hayes, G. D. Jones, H. F. MacDonald and C. T. Sennet. Proc. R. Soc. A289 (1966) 1.
[22] R. E. Schamu, W. M. Hartmann and E. Y. Yasaitis, Phys. Rev. 170 (1968) 822.
[23] G. O. Jones and J. M. Woddfine, Proc. Phys. Soc. 86 (1965) 101; W. M. Hartmann and R. J. Elliott, Proc. Phys. Soc. 91 (1967) 187.
[24] H. Dötsch, W. Gebhardt and C. H. Martius, Solid St. Commun. 3 (1965) 297.
[25] H. Dötsch, Phys. Stat. Solidi 31 (1969) 649; B. Fritz, in: Localized Excitations in Solids (op. cit.) p. 480.
[26] D. Bauerle and B. Fritz, Phys. Stat. Solidi 29 (1968) 639.
[27] H. Rosenstock and C. C. Klick, Phys. Rev. 119 (1960) 1198.
[28] R. Wallis and A. A. Maradudin, Prog. Theor. Phys. 24 (1960) 1055.
[29] G. S. Zavt, Sov. Phys. Sol. St. 5 (1963) 792.
[30] R. Fieschi, G. F. Nardelli and N. Terzi, Phys. Rev. A138 (1965) 203.
[31] T. Timusk, E. J. Woll Jr. and T. Gethins, Localized Excitations in Solids, ed. R. F. Wallis (Plenum Press, New York, 1968) p. 533.
[32] T. Gethins, Canad. J. Phys. 48 (1970) 580.
[33] W. Hayes and H. F. McDonald, Proc. R. Soc. A297 (1967) 503.
[34] B. Fritz, J. Gerlach and U. Gross, Localized Excitations in Solids (Plenum Press, New York, 1968) p. 504.
[35] W. Gebhardt and K. Maier, Phys. Stat. Solidi 8 (1965) 303.
[36] T. Timusk and M. V. Klein, Phys. Rev. 141 (1966) 664.
[37] Nguyen X. Xinh, Phys. Rev. 163 (1967) 896.
[38] J. B. Page and B. C. Dick, Phys. Rev. 163 (1967) 910.
[39] T. Gethins, T. Timusk and E. J. Woll Jr., Phys. Rev. 157 (1967) 744.
[40] R. W. McPherson and T. Timusk, Canad. J. Phys. 48 (1970) 2176.
[41] R. W. McPherson and T. Timusk, Canad. J. Phys. 48 (1970) 2917.
[42] M. De Souza, A. Diaa-Gongora, M. Aegerte and F. Lüty, Phys. Rev. Letters 25 (1970) 1426.
[43] W. Martienssen and H. Pick, Z. Phys. 135 (1953) 309.
[44] F. Lüty, J. Phys. C: Solid St. Phys. 28 (1967) 120.
[45] V. Narayanamurti, and R. O. Phol, Rev. Mod. Phys. 42 (1970) 201.
[46] B. Wedding and M. V. Klein Phys. Rev. 177 (1969) 1274.
[47] R. Capelleti, F. Fermi and R. Fieschi, Int. Conf. on Color Centers (Reading, 1971) Abstract 108 and private communication.
[48] D. Harrison and F. Lüty, Bull. Am. Phys. Soc. 12 (1967) 82; D. Harrison, Thesis, University of Utah (1969) unpublished.
[49] M. V. Klein, B. Wedding and M. A. Levine, Phys. Rev. 180 (1969) 902.
[50] A. F. Devonshire, Proc. R. Soc. A153 (1936) 601.
[51] P. Sauer, Z. Phys. 194 (1966) 360; P. Kuri and P. Sauer, Z. Phys. 194 (1966) 478.
[52] V. Narayanamurti, Phys. Rev. Letters 13 (1964) 693.
[53] H. S. Sack and M. C. Moriarty, Solid St. Commun. 3 (1965) 93.
[54] W. D. Seward and V. Narayanamurti, Phys. Rev. 148 (1966) 463.
[55] V. Narayanamurti, W. D. Seward and R. O. Pohl, Phys. Rev. 148 (1966) 481.
[56] A. Cabana, G. B. Savitsky and D. F. Hornig, J. Chem. Phys. 39 (1963) 2942.
[57] U. Durr and D. Baüerle, Z. Phys. 233 (1970) 94.
[58] D. Baüerle and B. Fritz, Solid St. Commun. 6 (1968) 353.
[59] A. J. Sievers, A. A. Maradudin and S. S. Jaswal, Phys. Rev. 138A (1965) 272.
[60] I. G. Nolt, R. A. Westwjg, R. W. Alexander Jr. and A. J. Sievers, Phys. Rev. 157 (1967) 730.

[61] A. A. Maradudin, Localized Excitation in Solids (op. cit.) p. 1.
[62] D. Bauerle and R. Hubner, Phys. Rev. **B2** (1970) 4252.
[63] A. J. Sievers, Elementary Excitations in Solids (op. cit.) p. 193.
[64] G. Benedek and E. Mulazzi, Phys. Rev. **179** (1969) 906.
[65] K. Thommen, Z. Phys. **186** (1965) 347.
[66] F. Lüty, S. Mascarenas and C. Ribeiro, Phys. Rev. **168** (1968) 1080.
[67] A. J. Sievers, Localized Excitations in Solids (op. cit) p. 34.
[68] C. De Jong, Solid St. Commun. **9** (1971) 527.
[69] R. D. Kirby, Phys. Rev. Letters **26** (1971) 512.
[70] G. Dolling, R. A. Cowley, C. Schittenhelm and I. M. Thorson, Phys. Rev. **147** (1966) 577.
[71] I. G. Nolt and A. J. Sievers, Phys. Rev. **174** (1968) 1004.
[72] H. F. McDonald and M. V. Klein, Localized Excitations in Solids (op. cit.) p. 46.
[73] R. Weber and P. Nette Phys. Letters **20** (1966) 493.
[74] A. J. Sievers, Phys. Rev. Letters **13** (1964) 310.
[75] R. D. Kirby, I. G. Nolt, R. W. Alexander Jr. and A. J. Sievers, Phys. Rev. **168** (1968) 1057.
[76] R. W. Alexander Jr., A. E. Hughes and A. J. Sievers, Phys. Rev. **B1** (1970) 1563.
[77] A. M. Kahan, M. Patterson and A. J. Sievers, Phys. Rev. **B14** (1976) 5422.
[78] D. B. Fitchen, Physics of Color Centers (op. cit.).
[79] F. C. Bauman and R. O. Pohl, Phys. Rev.
[80] R. D. Kirby, Phys. Rev. Letters **26** (1971) 512.
[81] R. O. Pohl, Localized Excitations in Solids (op. cit.) p. 434; Elementary Excitations in Solids (op. cit.) p. 259.
[82] R. F. Cadwell and M. V. Klein, Phys. Rev. **158** (1967) 851.
[83] F. C. Bauman and R. O. Pohl, Phys. Rev. **163** (1967) 843.
[84] R. D. Kirby, A. E. Hughes and A. J. Sievers, Phys. Rev. **B2** (1970) 481.
[85] A. J. Sievers and S. Takeno, Phys. Rev. **140A** (1965) 1030.
[86] P. P. Peressini, J. P. Harrison and R. O. Pohl, Phys. Rev. **100** (1969) 926.
[87] M. Gomez, S. P. Bowen and J. A. Krumhansl, Phys. Rev. **153** (1967) 1009.
[88] D. Walton, Localized Excitations in Solids (op. cit.) p. 395.
[89] G. Feher, I. Shepherd and K. B. Shore, Phys. Rev. Letters **16** (1966) 500; W. E. Bron and R. W. Dreyfus, Phys. Rev. **163** (1967) 304; T. L. Estle, Phys. Rev. **176** (1968) 1056; R. A. Herendeen and R. H. Silsbee, Phys. Rev. **188** (1969) 645; W. G. Von Holle, J. H. S. Wang, R. S. Scott and W. H. Flygare, Solid. St. Commun. **8** (1970) 1363; D. Blumenstock, R. Osswald and H. C. Wolf, Z. Phys. **231** (1970) 333; Phys. Stat. Solidi (**b)46** (1971) 217; A. V. Frantsesson, O. F. Dudnik and V. B. Kravchenko, Sov. Phys. Stat. **12** (1970) 126.
[90] M. C. Hetzler and D. Walton, Phys. Rev. Letters **24** (1970) 505.
[91] R. D. Kirby and A. J. Sievers, Phys. Letters **33A** (1970) 405.
[92] W. O. Wilson, R. D. Hatcher, G. J. Dienes and R. Smoluchowski, Phys. Rev. **161** (1967) 888.
[93] R. J. Quigley and T. P. Das, Phys. Rev. **164** (1965) 1185; Phys. Rev. **177** (1969) 1340.
[94] R. S. Scott and W. H. Flygare, Phys. Rev. **182** (1969) 445.
[95] J. G. Pescoe, W. R. Fenner and M. V. Klein, J. Chem. Phys. **60** (1974) 4208.
[96] R. T. Harley, J. P. Page and C. T. Wallner, Phys. Rev. **B3** (1971) 1365.
[97] O. Bimberg, W. Dultz, K. Fussgage and W. Gebhardt Z. Phys. **224** (1969) 364.
[98] I. Nair and C. T. Walker, Phys. Rev. **B3** (1971) 3446.
[99] W. Moller and R. Kaiser.
[100] J. P. Hurrel, S. P. S. Porto, T. C. Damen and S. Mascarenhas, Phys. Letters **26A** (1968) 194.
[101] V. S. Vasilov, U. S. Vinogradov, L. K. Vodopyanov and B. S. Vinarov in: Light Scattering in Solids, ed. M. Balkanski (Flammarion, Paris, 1971) p. 338.
[102] G. P. Montgomery and R. D. Kirby, in: Light Scattering in Solids, ed. M. Balkanski (op. cit.).

[103] M. Zigone, R. Beserman and M. Balkanski, in: Light Scattering in Solids, ed. M. Balkanski (op. cit.).
[104] W. Känzig, J. Phys. Chem. Solids 23 (1962) 479.
[105] J. A. Sussman, Phys. Kondens Materie 2 (1964) 146.
[106] R. Pirk, B. Zeks and P. Gosar, J. Phys. Chem. Solids 27 (1966) 1219.
[107] R. H. Silsbee, J. Phys. Chem. Solids 28 (1967) 2525.
[108] G. Pfister and W. Känzig, Phys. Kondens Materie 10 (1969) 231.
[109] R. H. Silsbee and I. Bojko J. Phys. Chem. Solids 34 (1973) 1971.
[110] N. E. Byer and H. S. Sack, J. Phys. Chem. Solids 29 (1968) 677.
[111] J. P. Harrisson, P. P. Peressini and R. O. Pohl, Phys. Rev. 171 (1968) 1037.
[112] H. Hartel, Phys. Stat. Solidi 42 (1970) 369.
[113] S. Kapphan and F. Lüty, Solid Stat. Commun 8 (1970) 349; Int. Conf. on Color Centers (Reading, 1971) abstract no. 68; B. G. Dick and O. Strauch, Phys. Rev. B2 (1970) 2200.
[114] F. Bridges, Crit. Rev. Solid St. Sci. 5 (1975) 1.
[115] F. Lüty, Phys. Rev. B10 (1974) 3667.
[116] H. Kinder, Phys. Rev. Letters 28 (1972) 1564; Z. Phys. 262 (1973) 295.
[117] R. Windheim and H. Kinder, Phys. Letters 51A (1975) 475.
[118] R. S. Scott and W. H. Flygare, Phys. Rev. 182 (1969) 445.
[119] F. Lüty, J. Phys. C: Solid St. Phys. 28 (1967) 125.
[120] I. Shepherd and G. Feher, Phys. Rev. Letters 15 (1965) 194.
[121] W. Känzig, H. R. Hart Jr. and S. Roberts, Phys. Rev. Letters 13 (1964) 543.
[122] A. T. Fiory, Phys. Rev. B4 (1971) 614.
[123] M. Gomez, S. P. Bowen and J. A. Krumhansl, Phys. Rev. 180 (1969) 926.
[124] W. N. Lawless, Phys. Chem. Solids. 30 (1969) 1161.
[125] R. D. Kirby, A. E. Hughes and A. J. Sievers, Phys. Rev. B2 (1971) 481.
[126] P. W. Anderson, Mat. Res. Bull. 5 (1970) 549.
[127] B. Fisher and M. W. Klein, Phys. Rev. Letters. 37 (1977) 756.
[128] G. Benedek, in: Physics of Impurity Centers in Crystals Proc. Int. Seminar on Selected Problems in the theory of impurity centers in crystals, (Tallinn, 1970, ed. G. S. Zavt).
[129] D. J. Channon, V. Narayanamurti and R. O. Pohl, Phys. Rev. Letters 22 (1969) 524.
[130] D. J. Channon, Thesis Cornell University, Ithaca, New York, (1970) unpublished.
[131] W. Heitler, in: Quantum theory of radiation, 3rd. ed. (Oxford, 1954) p. 196.
[132] J. H. Van Vleck, Phys. Rev. 59 (1941) 724.
[133] J. A. Giordmaine and F. R. Nash, Phys. Rev. 138A (1965) 1510; W. J. Brya, S. Geshwind and G. E. Deulin, Phys. Rev. Letters 21 (1968) 1800.
[134] T. R. Larson and R. H. Silsbee, Phys. Rev. B6 (1972) 3804; C. H. Anderson and E. S. Sabisky, Phys. Rev. Letters 21 (1968) 987; E. S. Sabisky and C. H. Anderson, Appl. Phys. Letters 13 (1968) 214.
[135] C. H. Anderson and E. S. Sabisky, Phys. Rev. 143 (1966) 223.
[136] R. D. Dynes, V. Narayanamurti, M. Chiu, Phys. Rev. Letters 26 (1971) 181.
[137] D. Walton, Phys. Rev. Letters 19 (1967) 305.
[138] I. B. Bersuker, Phys. Letters 20 (1966) 589; I. B. Bersuker and B. G. Vekhter, Sov. Phys. Solid Stat. 9 (1968) 2084; I. B. Bersuker, B. G. Vekhter, G. S. Danilchuck, L. S. Kremenchugskii, A. A. Muzalevskii and M. L. Rafalovitch, Sov. Phys. Solid Stat. 11 (1970) 1980; I. B. Bersuker, B. G. Vekhter and A. A. Muzalevskii, Ferroelectrics 6 (1974) 197.

3 | EFFECT OF ELECTRON-VIBRATION INTERACTION ON OPTICAL TRANSITIONS

3.1. Introduction

This chapter deals with the effect of electron–vibration interaction on the optical transitions associated with defect states. After discussing the basic theory of optical band shapes, we shall give some examples of defects featuring stronger or weaker coupling with the lattice. In this framework we shall discuss zero-phonon lines and associated vibrational structures. An important manifestation of electron–vibration coupling for orbitally degenerate electronic states (and in principle also for spin degeneracy, although in this case the associated splitting would be very small [29]) is the Jahn–Teller effect, which considerably modifies the optical response of the defect. Finally, we shall treat briefly the problem of defect-induced Raman scattering, with emphasis on resonance Raman scattering from F-centers; in the framework of resonance effects, a brief discussion about resonant scattering and the so-called hot luminescence will be presented, together with some experimental evidence.

3.2. Electron–vibration interaction and optical transitions

3.2.1. The adiabatic approximation and the Franck–Condon principle

The adiabatic approximation allows the separation of electronic and nuclear displacements (cf. chapter 2). Two equations, one in the nuclear coordinates, the other in the electronic coordinates, are the result of decoupling; these equations are coupled however by the electronic energy eigenvalue which acts as an effective potential of the nuclear equation of motion. The total wavefunction is written as:

$$\psi_n(\mathbf{r}, \mathbf{R}) = \varphi_l(\mathbf{r}, \mathbf{R}) \, \chi_{l,i}(\mathbf{R}),$$

where i is the vibrational quantum number and l describes the electronic state. It is important to remark that l is not a good quantum number for the vibrational wavefunction since it is associated with an electronic state which depends parametrically on the nuclear coordinate R. The ψ_n are an orthogonal set, because the electronic wavefunctions $\varphi_l(\mathbf{r}, \mathbf{R})$ are orthogonal; the vibrational wavefunction $\chi_{l,i}$ instead are not orthogonal:

$$\langle \chi_{l,i} | \chi_{l',i'} \rangle \neq 0. \tag{3.1}$$

The non-zero overlap of vibrational wavefunctions associated with different electronic states makes possible optical transitions which take place with the simultaneous creation (or destruction) of one or several vibrational quanta. In the harmonic approximation, the adiabatic potential may be written as:

$$W_l(\mathbf{R}) = W_l(\mathbf{R}_{0l}) + A\,(\mathbf{R} - \mathbf{R}_{0l}) + B\,(\mathbf{R} - \mathbf{R}_{0l})^2 \tag{3.2}$$

where the constants A and B depend on the electronic state, as does the equilibrium configuration $\mathbf{R}_{0l}$. The linear term in this equation must be zero (for the non-degenerate states we shall consider now) since $\mathbf{R}_{0l}$ is the equilibrium coordinate. Simplifying to one-dimensional space, the curve $W_l(\mathbf{R})$ will reduce to parabolas, and there will be one such parabola for each electronic state; the minima $\mathbf{R}_{0l}$ of each parabola will be mutually displaced, as can be seen from fig. 3.1. In one dimension, R is called the configuration coordinate. Configuration coordinate diagrams may also be used in a solid, if we assume coupling to only one totally symmetric dispersionless vibrational mode (the "breathing mode"). On this basis a simple theory of optical band shapes may be constructed which has interpreted a

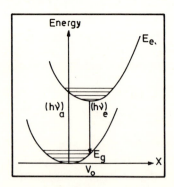

Fig. 3.1. Configuration coordinate diagram.

great number of defect optical properties, at least qualitatively [1]. The B parameters are connected to the force constants which appear in the dynamical matrix of the crystal (cf. chapter 2). In general, B depends on the electronic state, i.e. the curvature of the parabolas changes with electronic state. If B is assumed to be independent of the electronic state, we are in the linear coupling approximation. This is clear if we remark that the A parameter determines the coupling of the l electronic state with the crystal vibrations. If the electronic state does not influence the adiabatic potential the parabolas are not displaced (R_{0l} is the same for all l, and then $A_{l'} = A_l = 0$ by choosing correctly the coordinate): in this case the coupling is zero. If the adiabatic potential depends on the electronic state, but $B_l = B_{l'}$, A_l will be non-zero by taking for the coordinate system that of the state l. In the linear coupling approximation, the force constants, i.e. the vibrational frequencies, are independent of the electronic state. In this theory, one also neglects the dependence of the normal coordinate transformation on the electronic state. For a discussion of these and other approximations the reader is referred to the literature [2, 3].

In an optical transition, the defect passes from an initial state $n = (l, i)$ to a final state $n' = (m, f)$. The corresponding wavefunctions are:

$$\psi_n(\mathbf{r}, \mathbf{R}) = \varphi_l(\mathbf{r}, \mathbf{R}) \, \chi_{l,i}(\mathbf{R}),$$

$$\psi_{n'}(\mathbf{r}, \mathbf{R}) = \varphi_m(\mathbf{r}, \mathbf{R}) \, \chi_{m,f}(\mathbf{R}). \tag{3.3}$$

For electric dipole transitions, the transition probability is proportional to the square of $r_{n',n}$, defined by:

$$r_{n'n} = \langle \psi_{n'} | \mathbf{r} | \psi_n \rangle, \tag{3.4}$$

where

$$\mathbf{r} = -e \sum_\alpha \mathbf{r}_\alpha + e \sum_\beta Z_\beta \, \mathbf{r}_\beta, \tag{3.5}$$

where the indices α and β refer to electrons and nuclei respectively. Combining the last three equations and neglecting the nuclear terms which lead to IR absorption, we obtain:

$$r_{n'n} = \langle D_{ml}(\mathbf{R}) \, \chi_{mf}(\mathbf{R}) | \chi_{l,i}(\mathbf{R}) \rangle, \tag{3.6}$$

in which

$$D_{ml}(\mathbf{R}) = \langle \varphi_m | -e \sum_\alpha \mathbf{r}_\alpha | \varphi_l \rangle. \tag{3.7}$$

In the following chapter, we shall briefly discuss the complexity of determining the wavefunctions of an electronic state, even as simple as that associated with the F-center. Thus, when electronic states are not of direct interest, $D_{ml}(\mathbf{R})$ is not calculated from first principles but rather approximated heuristically. In particular, in the spirit of the adiabatic approximation, $D_{ml}(\mathbf{R})$ may be expanded about the equilibrium configuration $\mathbf{R}_{0l}$ of the initial electronic state:

$$D_{ml}(\mathbf{R}) = D_{ml}(\mathbf{R}_{0l}) + D_m^{(1)}\mathbf{R} + D_m^{(2)}\mathbf{R}^2. \tag{3.8}$$

The Condon approximation consists in assuming:

$$D_{ml}(\mathbf{R}) = D_{ml}(\mathbf{R}_0) \tag{3.9}$$

In this approximation the electronic matrix element is independent of the nuclear coordinate. The lattice has no time to move during the electronic transition: referring to fig. 3.1, the transition will be vertical, i.e. at constant $\mathbf{R}$. If D_{ml} is a constant, it may be taken out of the integral, yielding for the transition probability:

$$W_{li,mf} = |D_{ml}|^2 |\langle \chi_{mf} | \chi_{li} \rangle|^2. \tag{3.10}$$

In the Condon approximation, the effect of the electron–vibration coupling appears only in the vibrational overlap integral, which as we have stated will be generally non-zero since the vibrational wavefunctions χ corresponding to different electronic states are not orthogonal. It is important to note that the Condon approximation may not be used if $D_{ml}(\mathbf{R}_{0l}) = 0$. In this case in fact, the condition $D_{ml}(\mathbf{R}) = D_{ml}(\mathbf{R}_{0l})$ becomes too restrictive; in particular it leads to neglecting all vibration-induced transitions. Non-resonant Raman scattering is also intrinsically caused by a violation of the Condon approximation, since it is due to variations of $D_{ml}(\mathbf{R})$ caused by lattice variations. If $D_{ml}(\mathbf{R}) \neq 0$, the Condon integral [eq. (3.10)] will determine the optical band shapes, since the vibrational overlap depends strongly on the transition energy.

3.2.2. Theory of vibronic transitions

It is well known that in solids optical transitions are characterized by broad structureless bands or by sharp lines with associated broad band structures. This great diversity of behaviour may be interpreted in the framework of a single theory†. Most of the details of vibronic (vibrational + electronic) transitions may be understood on the basis of simple theory within the adiabatic, harmonic and Condon approximations. In the following we shall discuss these theories briefly following largely the excellent treatment by Rebane [5]. At first we shall treat the particularly simple and

---

† For a review, see the articles of Perlin [3] and McCumber [4].

practically interesting case of an electronic state interacting with a single high frequency localized vibrational mode of frequency ω_0. It is clear that this mode is caused in the lattice by the same defect associated with the electronic transition. The high frequency of the vibrational mode ensures that for most ordinary temperatures, only the ground vibrational state of the mode will be occupied. Thus the vibrational wavefunction is given by:

$$\chi_0(\xi_l) = K_0 \exp[-(\xi_l^2/2)], \tag{3.11}$$

where ξ_l, is the reduced coordinate

$$\xi_l = R_l(\hbar/m\omega_0)^{-1/2}. \tag{3.12}$$

It is convenient to translate the coordinate system so that the origin coincides with the equilibrium position in the excited electronic state ξ_{0m}, i.e. $\xi_l = \xi_{0m} + \xi_m$ where ξ_m is finite because in general there will be a finite displacement of the ions surrounding the defect when the electronic state is changed. Thus we have:

$$\chi_0(\xi_m) = K_0 \exp[-(\xi_0^2/2)] \exp[-(\xi_m^2/2)] \exp(-\xi_0\xi_m).$$

The vibrational wavefunction for the m excited electronic state in the vibrational state is:

$$\chi_f(\xi_m) = K_f \exp(\xi^2/2) \, (d^f/d\xi^f) \, \exp(\xi^{-2}),$$

with

$$K_f = (-1)^f/(2^f f! \sqrt{\pi})^{1/2}.$$

The overlap integral (Franck–Condon integral) of the vibrational function may be written as:

$$J_{0f} = K_0 K_f \exp(-\xi_0^2/2) \int_{-\infty}^{+\infty} \exp(-\xi_0\xi) \, d^f/d\xi^f \exp(-\xi^2) d\xi.$$

Integrating by parts f times we obtain:

$$J_{0f} = [(-\xi_0)^f \exp(-\xi_0^2/2)]/(2^f f!)^{1/2}.$$

Thus the transition probability for $l_0 \to mf$ is proportional to:

$$W_{0f} = \exp(-S_0) S_0^f/f!, \tag{3.13}$$

an expression in which:

$$S_0 = \xi_0^2/2 = m\omega_0^2 R_0^2/2\hbar\omega_0$$

is a dimensionless quantity connected with the energy difference between the minimum of the excited state parabola and the point on this parabola reached after a vertical transition from the groundstate equilibrium position.

This energy difference is just the energy dissipated into the lattice during the electronic transition through the lattice relaxation which accompanies it. This energy loss corresponds to the excitation of S_0 quanta of the localized vibrational mode and is called Stokes loss. This loss will increase with the coupling strength, i.e. when the lattice relaxation is larger. This basic result may be generalized to the case of phonons. However it is convenient to see what kind of spectrum will result from different values of S_0, which is the fundamental parameter for this problem (for the more general phonon case, S_0 is called the Huang–Rhys factor S).

For very weak coupling ($S_0 \sim 0.1$) there will be non-zero probability only for small f transitions. In the $S_0 = 0$ limit (no coupling) we would obtain only one line with unity intensity. For $S_0 = 0$, the parabolas for the various electronic states are not displaced and the vibrational wavefunctions of different electronic states will now be orthogonal: thus only transitions between states with the same vibrational quantum number will be allowed. The absorption line (which is actually a superposition of no quanta lines for the same electronic transition, a "quasi-line"), will be a purely electronic line and is called zero-phonon line in analogy with the case of interaction with phonons (or lack of it). For $S_0 = 0.1$ there are small sidebands to the electronic line (fig. 3.2). These additional lines correspond to transitions in which the vibrational quantum number changes by one or more units. This additional energy shifts to higher energy the optical transition. In the case of weak or intermediate coupling ($S_0 \sim 1$), the purely electronic line and the vibrational sidebands have comparable intensities (fig. 3.2): the multi-quantum transition probability has increased. Finally, in the case of strong coupling ($S_0 \sim 10$), the most probably transition is that with about ten

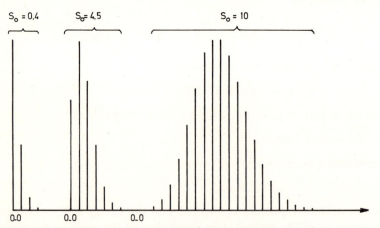

Fig. 3.2. Spectral distribution of absorption band shapes as a function of S_0.

quanta. The purely electronic line and the one or two quantum sidebands have negligible intensity. In fact, the purely electronic line has an intensity which is $\exp(-S_0)$ that of the total intensity. The structures shown in fig. 3.2, are clearly schematic and several effects may broaden the lines yielding partial or total overlap between the various lines; only the purely electronic line (within the approximation stated earlier) will not be broadened and in principle its width should be connected with the radiative lifetime of the excited state. In conclusion, we see that the interaction of an electronic state with a simple vibrational mode may lead to a great variety of absorption band shapes, from the atomic-like purely electronic line to a broad quasi-Gaussian distribution for the case of strong coupling.

We shall now consider the effects due to coupling with lattice phonons. In a perfect crystal, only zone-center phonons may couple to the optical transition, since the visible light wavevector is small relative to the extension of the Brillouin zone (BZ): thus the effect of vibrational dispersion is not seen. For a defect-related transition, the crystal translational symmetry is destroyed and the ensuing breaking of the wavevector conservation law causes from all points in the BZ to contribute to the optical transition. In general, the crystal vibrational modes will "modulate" the electronic state energy, yielding a continuous distribution for the optical band shapes instead of the delta-function structures obtained before for the localized mode. Furthermore, there will now exist very low frequency phonons for which the conditions $kT \ll \hbar\omega$ will not be valid, even at low temperatures. The optical absorption band shape will be given by the sum of all the final vibrational states averaged on all the initial vibrational states:

$$I(\Omega) = \text{Ave } (i) \sum_f' |\langle \chi_{mf} | \chi_{li} \rangle|^2, \tag{3.14}$$

where Ave (i) indicates the statistical average over all the initial states and $\sum'$ indicates that the sum is subject to the energy conservation law:

$$\hbar\Omega = E_{mf} - E_{li}.$$

For the groundstate the adiabatic potential may be written as:

$$E_l(q) = E_l^0 + \sum_k \tfrac{1}{2}\hbar\omega_k \, q_k^2 \tag{3.15}$$

and for the final state:

$$E_m(q) = E_m^0 + \sum_k \tfrac{1}{2}\hbar\omega_k \, q_k^2 + N^{-1/2} \sum_k A_k \hbar\omega_k \, q_k,$$

these expressions being valid in the linear coupling approximation. For each mode k, there is a dimensionless constant A_k which determines the coupling of the mode with the electronic state m. As before, this coupling will have the effect of displacing the equilibrium positions of the vibrational states in electronic state m relative to the corresponding positions in the state l.

This displacement is infinitesimal since each mode yields a contribution proportional to $N^{-1/2}$ (N number of degrees of freedom). However, this infinitesimal displacement may not be neglected since there are N modes which will contribute and will finally yield a finite displacement of the ions (in cartesian coordinates) after the defect has achieved the electronic transition. The solutions of the vibrational equation in the adiabatic approximation will be products of harmonic oscillator functions for all k modes. The band shape function may then be written as:

$$I(\Omega) = \text{Ave } (i) \sum_f' \Pi_k |\langle f_k | i_k \rangle|^2, \tag{3.16}$$

where f_k is the occupation number in mode k in the final state and i_k that for the initial state. Expression (3.16), was calculated by Pekar [6] and Huang–Rhys [7] with the simplifying hypothesis of dispersionless phonons. This method was then generalized by Pekar and Krivoglaz [8] to phonons with dispersion. The method is based on the fact that the equilibrium position displacement Δ_k after an electronic transition, is infinitesimal:

$$\Delta_k = -A_k/N^{1/2} \tag{3.17}$$

In the calculation only terms of the order of N^{-1} are kept: this means physically that only those transitions are considered in which each oscillator changes quantum number by at most one unit. In eq. (3.16) each mode is statistically independent and will yield a contribution proportional to its occupation number. Thus the statistical average may be evaluated by replacing each occupation number with its average:

$$\bar{n}_k = [\exp(\hbar\omega_k/kT) - 1]^{-1}; n = i, f.$$

For dispersionless phonons, the band-shape function becomes:

$$I(\Omega) = \exp\left[\sim S \coth\left(\frac{\hbar\omega}{2kT}\right) + P\left(\frac{\hbar\omega}{2kT}\right) \right] J_p\left(\frac{S}{\sinh(\hbar\omega/rT)}\right), \tag{3.18}$$

where

$$S = \frac{1}{2}\sum_k \Delta_k^2 = \frac{1}{2N}\sum_k A_k^2 \tag{3.19}$$

is called the Huang–Rhys factor and P is an integer defined by:

$$\hbar\Omega - E_{lm}^0 = P\hbar\omega, \tag{3.20}$$

ω being the phonon frequency and E_{lm}^0 the purely electronic transition energy. Eq. (3.20), is just the energy conservation law. $J_p(x)$ is the modified Bessel function. It is interesting to see that (3.18) reduces to the distribution

(3.13), obtained for a localized mode in the $T \to 0$ limit. At high temperatures $I(\Omega)$ reduces to a Gaussian:

$$I(\Omega) = I_0 \exp \left(- \frac{\hbar(\Omega - \Omega_0)^2}{kT} \cdot \frac{1}{\omega S} \right)$$

with halfwidth given by:

$$\Delta\Omega = 2\omega[2 \log(S_k T/\hbar\omega)^{1/2}] \, T^{1/2}.$$

The hypothesis of dispersionless phonons may be eliminated by allowing p to vary continuously. The series of lines of fig. 3.2, then becomes a continuous band whose shape will vary according to the value of the Huang–Rhys factor following the scheme of fig. 3.2.

3.2.3. Zero-phonon lines and vibrational sidebands

For strong coupling ($S \sim 10$), the optical bands are broad and structureless and it is thus difficult to extract from eq. (3.18) information about the coupling and the phonons which interact with the defect. The spectra are more interesting for weak coupling ($S \sim 1$); in this case (fig. 3.2) the most probable transitions are those corresponding to the purely electronic line and the one-phonon sideband. We shall now study in more detail the behavior of the zero-phonon line and in particular its variation with temperature. At very low temperature, only the groundstate vibrational level will be occupied; let us call it 0. The zero-phonon transition will take place from this level to the ground vibrational level of the excited electronic state, O'. If temperature increases the first excited vibrational level of the ground electronic state will begin to be populated, and there will be two other possible zero-phonon lines corresponding to the $0 \to 0'$ and $1 \to 1'$ transitions respectively. As the temperature increases further, there will be contributions from the $2 \to 2'$ transitions, etc, (the contributions will be degenerate only in the linear coupling approximation of course). The zero-phonon line will then be the sum of all these transitions. The temperature variations of the zero-phonon line may be calculated from the preceding equations. For this, the excited state vibrational wavefunction is expanded in a Taylor series about the equilibrium position relative to the initial state. In the linear coupling approximation:

$$\chi_{mf}(q_k) = \chi_{lf}(q_k - \Delta_k),$$

where Δ_k, is the displacement associated to the K mode. We obtain:

$$\chi_{mf}(q_k) = \chi_{lf}(q_k) - \frac{d\chi_{lf}}{dq_k}\Delta_k + \frac{1}{2}\frac{d^2\chi_{lf}}{dq_k^2}\Delta_k^2 + \dots$$

Stopping the expansion at the quadratic term, we obtain for the intensity of the zero-phonon line:

$$I(T) \sim \exp\left(-\sum_{k=1}^{N} 2 S_k \frac{kT_k}{\hbar\omega_k}\right), \tag{3.21}$$

where the reduced temperature T_k is given by:

$$kT_k = \tfrac{1}{2}\hbar\omega_k \coth\left(\hbar\omega/2T\right). \tag{3.22}$$

For $T \to 0$, $kT_k = \tfrac{1}{2}\hbar\omega_k$ and;

$$I(T) \sim \exp\left(-\sum_k S_k\right) = \exp(-S),$$

an expression which yields the zero-phonon line intensity (relative to the total intensity of the distribution). At high temperature $(kT \gg \hbar\omega_{10})$, $T_k \to T$ and thus:

$$I(T) \sim \exp\left(-kT \sum_k \frac{2S_k}{\hbar\omega_k}\right). \tag{3.23}$$

The zero-phonon line intensity decreases exponentially with increasing temperature and its variation may be described by a "Debye–Waller" factor [eq. (3.23)]. Since the total intensity of the band is constant in the Condon approximation, the diminution of the zero-phonon line will result in an increase of the multiple phonon transition continuum. The zero-phonon line is analogous to the Mössbauer line [9]; their temperature variations are the same since the Mössbauer line corresponds to the recoilless emission of a gamma ray from the nucleus (i.e. without phonon emission). In fact, the analogy cannot be carried too far. For instance, the width of the zero-phonon line should be radiation limited since, by definition no phonon (in first order) participates in the transition. There should be then zero-phonon lines with narrower widths than atomic lines in gases which are always broadened at least by the Doppler effect. Unfortunately, crystal inhomogeneities and internal stresses considerably broaden the zero-phonon line, and this is not the case for the Mössbauer line. In any case this broadening may be used to get information about the state of internal stresses in the crystal [10].

For intermediate coupling, the zero-phonon line is accompanied by multiple-phonon sidebands. For one-phonon sidebands, $f = i + 1$, the preceding formalism may be used to calculate the band shape:

$$\chi_{mf}(q_k) = \chi_{m,i+1}(q_k) = \chi_{l,i+1}(q_k - \Delta_k),$$

expanding and using the properties of the $\chi(q)$ functions, we obtain the overlap integral:

$$J_{i,i+1} = -\Delta_k[(i + l)(\omega_k/2\hbar)]^{1/2}$$

and the transition probability will be proportional to:

$$W^k_{i,i+1} = \frac{i + 1}{2\hbar}\Delta^2_k\omega_k = \frac{i + 1}{2\hbar}\frac{A^2_k}{N}\omega_k.$$

The transition probability for N-mode transitions will be:

$$W_{i,i+1}(\omega) \sim \frac{1}{N}\sum_k A^2_k\omega_k$$

and since ω_k is continuous, the sum reduces to an integral, yielding:

$$W_{i,i+1}(\omega) \sim A^2(\omega)\rho(\omega) \tag{3.24}$$

where $\rho(\omega)$ is the phonon density of states and $A^2(\omega)$ the effective coupling given by the average of A^2_k over all the modes of frequency ω. The one-phonon sideband, reflecting both the phonon density of states and the electron–phonon coupling, may have peaks originating from maxima in the phonon density of states, or from maxima in $A(\omega)$ indicating a preferential coupling of the defect electronic states to some given vibrational modes.

In this section we have considered only absorption processes. A completely analogous line of reasoning may be followed for emission processes. It is evident from fig. 3.1, that the absorption and emission spectra for transitions between non-degenerate electronic states will be identical and have mirror symmetry relative to the zero-phonon line (always in the linear coupling approximation). If one of the electronic states is generated, the situation becomes more complicated due to the Jahn–Teller effect and the mirror symmetry will be broken.

3.3. Vibronic spectra

3.3.1. F-center aggregates

The F_2-center is found in three charge states: F_2, F_2^+, F_2^-. In fig. 3.3, we show the absorption spectra of these centers at 4.2 K in LiF [11, 26]. These defects feature all possibilities of band shapes: F_2 has no structure, F_2^+ a weak zero-phonon line with some accompanying structure, and F_2^- a strong zero-phonon line and strong well defined one-phonon sideband. From the preceding discussion we see then that F_2^- and F_2 are examples of weak and

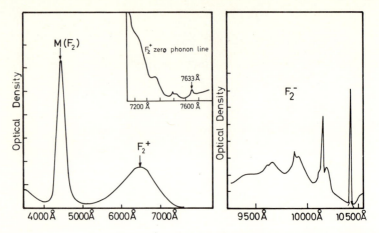

Fig. 3.3. Absorption spectra of F_2, F_2^+ and F_2^- in LiF (after Farge et al. [11] and Fetterman [26]).

strong coupling respectively. A similar set is given by the F_3, F_3^+ and F_3^- centers respectively [12]. In some cases there is mirror symmetry between absorption and emission; in other cases there are important differences which may be due either to a strong quadratic coupling or to the Jahn–Teller effects for transitions involving degenerate electronic states.

It is not easy to interpret these spectra. In principle, for weak coupling, the interpretation is easier since the optical bands are well resolved. Nevertheless there are still a certain number of difficulties: transitions involving one or two phonons may be degenerate; there may be important contributions from localized or resonant modes associated with the defect; finally, the dependence of the coupling on the different phonon types may modify considerably the one-phonon sideband shape. In the ideal case in which none of these complications existed the one-phonon sideband would just reproduce the one-phonon density of states of the pure crystal. In fig. 3.4, we may compare such density of states evaluated by inelastic neutron scattering measurements in LiF [13] with the one-phonon sidebands for the F_2^-, F_2^+ and F_3^- centers. The best agreement is obtained for the F_2^+-center, which is also the more strongly coupled defect. In this case also however, the agreement stops when intensities are considered which means that the coupling depends on the different types of phonons involved. It is not easy to assign the most resolved peaks to maxima in the phonon density of states; in general it is postulated that they originate from resonant modes. Such an hypothesis is not evident, at least if it is not verified by some other method. For F-aggregate centers in particular, there are always strong concentrations of F-centers which make IR or thermal conductivity measurements practically impossible. Some progress was made using

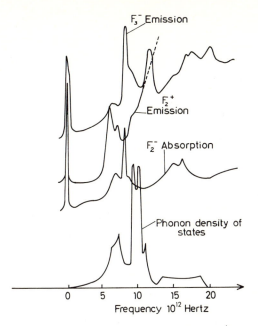

Fig. 3.4. Comparison of one-phonon emission sidebands for F_2^-, F_2^+ and F_3^- in LiF with the one-phonon density of states obtained by inelastic neutron scattering (after Fitchen [11] and Farge and Fontana [14]).

resonant Raman spectroscopy. Using the resonance enhancement (see the last section of this chapter) the contribution from the defect of interest may be isolated; for the F_2^+-center, the 6328 Å line of the He–Ne laser is very close to the maximum of the absorption band. In fig. 3.5, we have shown the resonant Raman scattering on F_2^+ in LiF, which should reflect to some extent the vibrational density of states (see chapter 2), compared to the one-phonon emission sideband [14]. The agreement between the spectra is reasonably good, which confirms that at least for F_2^+ the one-phonon sideband is due basically to the crystal phonon only. The interpretation of vibronic spectra is still on a qualitative level. In any case, an important step has been made with the definition of the symmetry selection rules for the phonon participation with the vibrational structure. For orbitally non-degenerate states, only the totally symmetric mode (in the symmetry group of the defect) will contribute. If the defect has a center of symmetry that is located at a lattice site, only even modes will contribute [15]. For F-center aggregates this rule does not hold since the center of symmetry is not located at a lattice site [16].

In conclusion we report in table 3.1 the positions of several typical zero-phonon lines in ionic crystals.

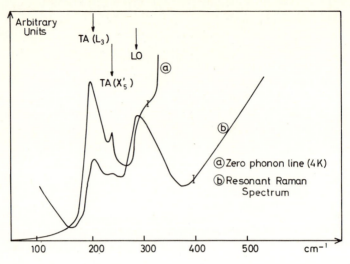

Fig. 3.5. The resonant Raman spectrum and the one-phonon emission sideband of F_2^+-center in LiF (after Farge and Fontana [14]).

Table 3.1
Zero-phonon lines of F-aggregate centers in some crystals.

Center	LiF	NaF	NaCl	KCl	KBr	CaF$_2$	SrF$_2$	CaO
F$^+$								3557[a]
F$_2$						5396[b]	6053[b]	
						⟨110⟩[g]	⟨110⟩[g]	
F$_2^+$	7633[c]		13060[c]	15262[c]				
F$_2^-$	10404[c]		15490[d]	20150[c]				
F$_3$		3909[c]	4474[c]	6329[c]	7417[c]	8163[c]	6774[e]	6792[f]
F$_3^+$	4874[c]	5456[c]		10218[c]				
F$_3^-$	8328[c]	10721[c]		16295[c]	17705[c]	7327[f]	8350[c]	
N$_1$(F$_4$?)	5234[c]	5803[c]	8373[c]	9924[c]	10978[c]			
N$_2$(F$_4$?)			8681[c]	10331[c]	11338[c]			

[a] J. C. Kemp, W. M. Ziniker, J. A. Glaze and J. C. Cheng, Phys. Rev. **171** (1968) 1024.
[b] J. H. Beaumont, A. L. Harmer, W. Hayes, J. Phys. C: Solid St. Phys. **5** (1972) 1475.
[c] From table 5.1 of the review article by D. B. Fitchen in: Physics of Color Centers (op. cit.) p. 306.
[d] I. Schneider and C. E. Bailey, Solid St. Commun. **7** (1969) 657.
[e] J. H. Beaumont, A. L. Harmer and W. Hayes, J. Phys. C: Solid St. Phys. **5** (1972) 257.
[f] J. H. Beaumont, A. L. Harmer and W. Hayes, J. Phys. C: Solid St. Phys. **5** (1972) 275.
[g] In these crystals, F$_2$-centers can be oriented along ⟨100⟩ or ⟨110⟩.

3.3.2. Molecular impurities

Until now, we have discussed separately the interaction with localized modes and with phonons. This distinction is of course artificial since a defect which interacts with a localized mode will also interact with lattice phonons. This is particularly evident for molecular impurities, which have their own intrinsic degrees of freedom. In general, the molecular vibrations are sufficiently decoupled from the host crystal ones that one may treat separately their coupling with the electrons of the system. In general, the molecular localized modes have much higher energy than the lattice phonons, and thus the vibronic structure will resemble the distributions of fig. 3.2. If, now, the interaction with phonons is turned on, each localized mode vibrational line will be substituted by a sideband structure characteristic of the electron–phonon coupling. Since the zero moment of each line must remain unchanged the area of the original Lorentzian will be redistributed on the structured sideband characteristic of the coupling with lattice phonons. This structure will replicate for the various lines of the localized mode distribution. The emission of $KBr:O_2$ at low temperatures is a good example of this situation [17] (see fig. 3.6). In this case, the electron-phonon interaction is sufficiently weak that only the one-phonon sideband structure is present. The replicas are well separated since the localized mode frequency is $1100 \ cm^{-1}$ whereas the LO phonon frequency maximum is $167 \ cm^{-1}$. For each of these replicas the "zero-phonon" line corresponds to the excitation of the localized mode without absorption or emission of phonons.

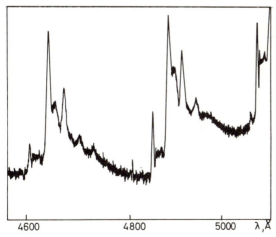

Fig. 3.6. Emission spectra of $KBr:O_2^-$ at 4.2 K (after Rebane ref. [17 p. 8).

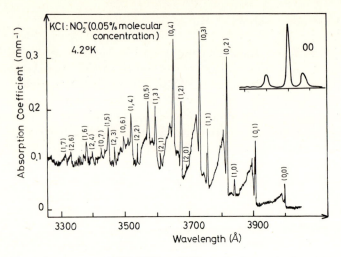

Fig. 3.7. Absorption spectra of $KCl:NO_2^-$ at 4.2 K. In the insert, the fine structure of the zero-phonon line (after Evans et al. [19]).

The NO_2^- impurity in alkali-halides is another interesting case [18, 19]. In fig. 3.7, we show a low temperature absorption spectrum of $KCl:NO_2^-$. Here the phonon-coupling replicas of the localized mode structure may also be observed. In this case, the situation is a bit more complicated, since the molecular ion has a more complex structure than O_2^-: in addition to the internal vibrations there are also librational modes [19, 20, 21]. These modes influence the absorption spectrum by causing a resolved fine structure in the zero-phonon lines of this distribution [19] (see insert in fig. 3.7). Since the molecule also has very low frequency rotational modes, it is possible to align it at very low temperature by applying uniaxial stress (cf. chapter 2). This molecular alignment yields a linear dichroism for the transitions of NO_2^- [18, 22]. Emission spectra yield similar results. With measurements of this type complemented by IR, Raman and thermal conductivity measurements, it may be shown that the substitutional NO_2^- movements can be classed in two categories:

internal molecular vibrations which are perturbed by the environment and which will be similar in the different alkali-halides;

rotational and librational modes which will be very sensitive to the surroundings.

3.3.3. Piezospectroscopic effects

All the defects studied up to now have internal structure, and have internal symmetry which is different from that of the crystal. The F_2-center for instance has axial symmetry (D_{2h}) in a cubic crystal (O_h), with the symmetry axis along the $\langle 110 \rangle$ directions. In a cubic crystal there are six equivalent $\langle 110 \rangle$ directions and thus the F_2-center will have a sixfold "orientational" degeneracy. More generally, if the point group of the crystal is of order G and the symmetry group of the defect (which must be a subgroup of the crystal one) is of order g, then the order r of orientational degeneracy will be G/g. It is important to notice that such degeneracy adds to possible electronic degeneracies of the defect. For the particular case of F_2, the electronic states cannot be degenerate since D_{2h} has no degenerate irreducible representations; in this system there will be only orientational degeneracy.

For the F_3-center (symmetry C_{3v}) there will be both orientational and electronic degeneracy. For simplicity we shall consider only the first case (orientational degeneracy only). Such degeneracy may be lifted by an external perturbation such as an electric field or uniaxial stress. Kaplyianskii has developed a complete theory of these effects [23]. Without stress, the optical transition energy for the electronic states l and m is independent of the orientation of the defect. This energy will be shifted by Δ under the effect of an applied stress, where Δ is given by; (in the linear coupling approximation):

$$\Delta = A_{xx}\sigma_{xx} + A_{yy}\sigma_{yy} + A_{zz}\sigma_{zz} + 2A_{yz}\sigma_{yz} + 2A_{zx}\sigma_{zx} + 2A_{xy}\sigma_{xy}, \quad (3.25)$$

where $\{\sigma_{ik}\}$ is the stress tensor and the coefficients $\{A_{ik}\}$ form a second-order symmetric tensor $\{A_{ik}\}$ called the piezospectroscopic tensor by Kaplyianskii. In fig. 3.8, (taken from Kaplyianskii's work) we show the polyhedra associated with the various possible symmetries of the defects, the possible orientations of dipolar transitions, and the corresponding forms of the $\{A_{ik}\}$ tensor. The x, y, z axes are just the cubic crystallographic axes. Three parameters yield an empirical description of the piezospectroscopic effect:

the number of components in which the optical line is split;

the splitting energy;

the polarization of the components.

The number of components is determined by the number of defect orientations which are not equivalent any more relative to the stress direction. For instance, there will be two components for the F_2-center with a [110] stress. The splitting magnitude may be calculated for each case from eq. (3.25). For F_2-centers and [100] stress, this splitting will be A_{11}, A_{22}, A_{33} depending on the relative orientation of the defect and stress. Finally, the applied stress will induce a linear dichroism since the components of the line will absorb only for well defined polarizations of the incident light.

Experimentally, it is the measurement of the stress-induced linear dichroism that allows detection of the piezospectroscopic effect. In general, the

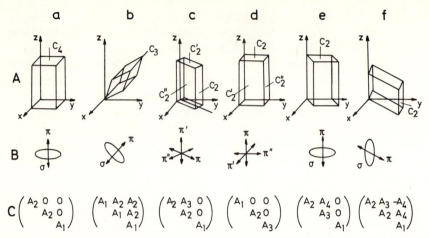

Fig. 3.8. Schematic representation of anisotropic centers in a cubic lattice. (a) Polyhedra of symmetry equivalent to that of the defect; (b) the center dipolar oscillators; (c) the piezo-spectroscopic tensors for each center after Kaplyanskii [23].

splitting is weak and thus it is easier to determine for narrow lines such as zero-phonon lines. Most studies thus have been carried out on weakly coupled defects, and particularly on some F-center aggregates. This effect however, has been observed for a broad band transition [25]. The piezo-spectroscopic effect of F_2^- in LiF is shown in fig. 3.9 [26]. The effect is

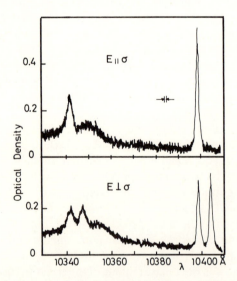

Fig. 3.9. Absorption spectrum of the F_2^--center in LiF for [100] stress (after Fetterman [26]).

proportional to the applied stress, and this allows the determination of A_{11}, A_{22}, A_{33}, by measuring the effect with the stress in the three directions [100, 110, 111]. The measurement of these coefficients is important since they are directly related to the elastic coupling of the defect with the lattice, and in particular to the coupling of the involved electronic states with the long wavelength, totally symmetric acoustic phonons. This type of measurement also allows the determination of the defect symmetry.

We must note however that each symmetry class contains several point groups. For instance, in the tetragonal class there are the C_{4v}, C_{4h}, D_{4h}, D_4 and S_4 groups. The piezospectroscopic measurements do not allow a discrimination among these groups, and thus complementary measurements are necessary: the Stark effect [27], polarized optical bleaching, or defect reorientation under uniaxial stress or electric field. These techniques have been briefly discussed in chapter 1.

3.4. Optical transitions between degenerate electronic states

Up to now, we have considered the electron–phonon coupling only in the case of non-degenerate electronic states. In the several approximations made, the main effect of the coupling is the displacement of the equilibrium positions of the nuclei following an electronic transition. For degenerate electronic states, the matrix element of the coupling to non-totally symmetric phonons will not be necessarily zero. This coupling may lift partially or totally the electronic degeneracy and mix the relevant electronic wavefunctions. In this case the electron–phonon coupling may strongly modify the nature of the electronic state. Furthermore the validity of the adiabatic approximation becomes questionable. Let us assume in fact that two electronic states are separated by an energy $E_0 \gg \hbar\omega_{\text{Debye}}$; the adiabatic approximation is valid and electronic and vibrational coordinates may be separated. If now E_0 diminishes and approaches zero, at a certain time it will become of the same order of magnitude as phonon or local mode energies, i.e. the basic criterion for the validity of the adiabatic approximation is not satisfied. This is all the more applicable to the case of $E_0 = 0$, i.e. degenerate electronic states. The electron–phonon coupling will then considerably influence degenerate electronic states: the ensemble of the various effects is called the Jahn–Teller effect.

3.4.1. The Jahn–Teller effect

Let us consider a molecule with a degenerate electronic state. As mentioned, in this case the state is linearly coupled to the non-totally symmetric vibrational coordinates; such linear coupling was first considered in the classic paper by Jahn and Teller [28] and their theorem may be enunciated as follows: "each non-linear molecule possessing spin or orbitally degenerate electronic state is unstable against a distortion which tends to lower its symmetry and (consequently) to lift the electronic degeneracy". The only exception to the validity of the theorem is Kramer's spin degeneracy which cannot be lifted by any electric potential, but Jahn showed that the effect is also practically unobservable in cases of usual spin degeneracy. Although the theorem so stated is certainly correct, a deeper understanding of its true significance with respect to a fully quantum mechanical treatment of the system was achieved only after the works of Moffitt and co-workers [30], Longuett-Higgins et al. [31], and Ham [32]. It is important to note that the theorem in its present formulation can be easily applied to localized defects in crystals if we make use of the quasi-molecular model for such systems; this consists in considering the defect or impurity center plus its nearest neighbors as a molecule, taking into account the effect of lattice vibrations in some average sense, consisting simply in a redefinition of the normal coordinate of the quasi-molecule. The theoretical justification for such procedure is provided by the work of Toyozawa and Inoue [33] who showed that, as far as the potential energy is concerned, such a procedure is exact. Given the enormous simplification that can be achieved in this way, we shall use the quasi-molecular model from now on. Let us review very briefly the demonstration of the J–T theorem. The normal coordinates q_k of the (quasi) molecule span the irreducible representations of the symmetry group of the system; if V_0 is the potential energy of the undistorted system (all $q_k = 0$), the energy of a generic distorted configuration can be written as:

$$V(Q) = V_0 + \sum_k \frac{\partial V}{\partial q_k} q_k + \frac{1}{2} \sum_k \sum_{k'} \frac{\partial^2 V}{\partial q_k \partial q_{k'}} q_k q_{k'} + \dots, \qquad (3.26)$$

where the dependence of V on the electronic coordinates has been omitted for simplicity. Consider now a generic electronic state, $|i\rangle$. The expectation value of V in such a state is:

$$\langle i| V(Q)|i\rangle = \langle i| V_0|i\rangle + \langle i| \partial V/\partial q_k|i\rangle q_k + \dots. \qquad (3.27)$$

Now, if $\langle i| \partial V/\partial q_k|i\rangle = a \neq 0$, the undistorted configuration is unstable because if $a > 0$, for instance for small $q_k < 0$ the left-hand side of (3.27) is smaller than $\langle i| V_0|i\rangle$. According to a well known theorem of elementary group theory, the matrix element

$$F_k = \langle i| \partial V/\partial q_k|i\rangle$$

is different from zero if the symmetric produce $[\Gamma_i]^2$ contains Γ_k where Γ_i and Γ_k are the irreducible representations according to which $|i\rangle$ and q_k

transform respectively. By examining all possible irreducible representations of point groups, Jahn–Teller found that if $|i\rangle$ is degenerate $[\Gamma_i]^2 \in \Gamma_k$ (Γ_k non-totally symmetric) except when the molecule is linear. Thus the system tends to lower its symmetry and consequently to lift the electronic degeneracy. But what is precisely meant by the locution "the system *tends to lower its symmetry*"? Actually, the Hamiltonian (3.26), like any observable physical quantity, spans the totally symmetrical irreducible representation Γ_1 of the group at the system, and as such it cannot lower the symmetry, just as a hydrostatic pressure cannot change the shape of a sphere. In fact, we shall see in more detail in the next section that the Jahn–Teller interaction always gives rise to (at least) as many distorted stable configurations, as is the initial degree of electronic degeneracy, and that there are always vibrations which allow the system to tunnel among these configurations and to restore the apparently broken symmetry. On the other hand, if the J–T stabilization energy E_{JT} (see next section for a more precise definition of this quantity) is large enough as compared to the vibrational quantum $\hbar\omega$, the tunnelling rate among different distorted configurations may be very small, such that during the characteristic time of a measurement the system appears as being distorted. Although in this case most spectroscopic manifestations of E_{JT} are as if the symmetry of the system were actually lowered (for instance, linear polarization of the emitted light if the J–T active state is the excited electronic state), this only means that the system is observed in a non-stationary state. In fact, if we observed the system for a long enough time Δt such that ΔE was small enough to descriminate between the ground and excited *vibronic* states, the unsymmetric features (e.g. polarization) would disappear, in agreement with the above statement that the groundstate must have the same symmetry as the Hamiltonian.

When the Jahn–Teller energy and $\hbar\omega$ are comparable, no unsymmetrical feature is observed: yet, the E_{JT} has observable consequences, consisting of a quenching of some orbital operators. We shall consider such quenching later. The first situation ($E_{JT} \gg \hbar\omega$) is usually referred to as the "static JTE", while when $E_{JT} \sim \hbar\omega$ the effect is usually called "dynamical". This is probably a bad nomenclature [34]. In fact, as we shall see, the method by which J–T problems are usually treated is a sort of adiabatic approximation, modified to be applicable to degenerate electronic states [31], in which the total (vibronic) wavefunctions are linear combinations of the form:

$$\Psi_m = \sum a_{mn} \varphi_n(\mathbf{q},\mathbf{Q}) \, \chi_n(Q) \tag{3.28}$$

It seems thus more appropriate to call the "static J–T problem" the first step of the calculation where, as in the usual adiabatic approximation, one solves for electronic eigenvalues and eigenfunctions alone, and call the "dynamic J–T problem" the whole calculation including nuclear motion.

3.4.2. *Jahn-Teller Effect on an orbital triplet T, in cubic (O_h) symmetry*

Let us consider an impurity or defect at a site of octahedral symmetry with a triply degenerate (excited) state T, such as on F-center or a Tl^+-like impurity in alkali-halides. According to group theory, a T_1 (or T_2) electronic state is coupled to three kinds of vibrational coordinates by the JTE, because:

$$[T_1]^2 = [T_2]^2 = \alpha_{1g} + \varepsilon_g + \tau_{2g},$$

(irreducible representations relative to vibrations are indicated by Greek letters). The α_{1g} mode is unimportant because it is totally symmetrical and all it can do is to shift all degenerate levels by the same amount. The ε_g and τ_{2g} modes are J–T active; this problem is accordingly called the $T \otimes (\varepsilon_g + \tau_{2g})$ J–T problem, and has been one of the most widely studied due to the large number of physical systems it concerns. The static problem was first considered by Van Vleck [35] and more extensively by Opik and Pryce [36]. As is well known the three components of T transform like the cartesian coordinates and are usually indicated as $|x\rangle, |y\rangle, |z\rangle$. If K_ε and K_τ are the elastic constants for ε_g and τ_{2g} modes respectively (we assume for the moment elastic restoring forces) it can be easily seen by group theoretical arguments [34, 37] that the Hamiltonian in matrix form, within the basis $|x\rangle, |y\rangle, |z\rangle$, is given by:

$$H = \begin{array}{c|ccc} & |x\rangle & |y\rangle & |z\rangle \\ \hline |x\rangle & -b(q_2 - q_3/3^{1/2}) & cq_6 & cq_5 \\ |y\rangle & cq_6 & b(q_2 + q_3/3^{1/2}) & cq_3 \\ |z\rangle & cq_5 & cq_4 & -2bq_3/3^{1/2} \end{array}$$

$$+ [K_\varepsilon(q_2^2 + q_3^2) + K_\tau(q_4^2 + q_5^2 + q_6^2)], \qquad (3.29)$$

where $b = \langle x | \partial V/\partial q_3 | y \rangle$; $c = \langle x | \partial V/\partial q_6 | y \rangle$, q_2 and q_3 are the ε_g coordinates; q_4, q_5, q_6 are the τ_{2g} coordinates, and I is a 3×3 unit matrix. The secular equation corresponding to (3.29) is cubic, and although a semi-analytical solution is possible in some cases [33] it is far simpler just to find the energies and symmetries of the stationary points of the five-dimensional surface $E(q_2, \ldots, q_6)$, and then investigate which of such points are stable (=minima) and under which conditions. Such a detailed analysis has been carried out by Opik and Pryce [36] and gave the following results. There are three kinds of stationary points: (a) three equi-energetic points corresponding to a tetragonal distortions of the octahedron along the coordinate axes x, y, z; these are minima if and only if $b^2/K_\varepsilon > c^2/K_\tau$; (b) four equi-energetic points corresponding to trigonal distortions of the octahedron along the $\langle 111 \rangle$ crystallographic axes, these are minima if and only if $c^2/K_\tau > b^2/K_\varepsilon$; (c) six stationary points of intermediate (orthorhombic) symmetry, corresponding to distortions along the $(\pm 1 |, 0, 0)$ axes which can never be minima.

Thus, in the present approximation [i.e. harmonic restoring forces and only linear terms of Hamiltonian (3.29)] tetragonal and trigonal minima do not coexist and orthorhombic minima do not exist. Such a dichotomy turns out to be just the result of the mentioned approximations; actually, inclusion of both quadratic terms in Hamiltonian (3.29) the so-called quadratic Jahn–Teller effect [38–40] and/or anharmonicity [41], allows orthorhombic minima and the coexistence of minima of different symmetry. Also, stationary points of lower symmetry appear which can never be minima [39].

The role of spin–orbit interaction is also rather important in determining the solutions of the static problem [36, 37, 40, 42], but we cannot discuss it in detail here. Some aspects of particular concern about impurities will be presented in the next chapter. Let us now turn to the dynamical problem. The whole $T \otimes (\varepsilon + \tau)$ dynamical problem is too difficult to be solved exactly but we have a variety of particular or perturbative solutions which, put together, provide rather definite answers to many questions, at least from a qualitative point of view.

The first thing to do is to put $c = 0$ in the Hamiltonian (3.29): in this case the dynamical problem becomes trivial. In fact, with $c = 0$, the potential surface is easily seen to be constituted by a three-dimensional paraboloid centered at $q_{4,5,6} = 0$ in the τ_{2g}-subspace, and by three two-dimensional, uncoupled, paraboloids (see fig. 3.10) in the ε_g-subspace. It is then clear that the vibronic eigenvalues at eigenfunctions are (three times) those of a displaced harmonic oscillator and it is a very straightforward matter [32] to write them down and to realize that the ground vibronic state is a triplet $|T\rangle$, the components of which are:

$$|T_x\rangle = |x\rangle|u_1\rangle|\varphi_0\rangle$$
$$|T_y\rangle = |y\rangle|u_2\rangle|\varphi_0\rangle \qquad\qquad (3.30)$$
$$|T_z\rangle = |z\rangle|u_3\rangle|\varphi_0\rangle$$

where $|x\rangle$, $|y\rangle$, $|z\rangle$ are the electronic states, $|u_{1,2,3}\rangle$ the $n_\varepsilon = 0$ vibrational wavefunctions centered on the respective paraboloids and $|\varphi_0\rangle$ the $n_\tau = 0$ vibrational wavefunction centered on the origin of the τ_{2g}-subspace. Although the derivation of (3.30) is trivial, their qualitative meaning (i.e.

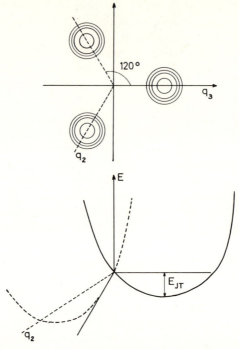

Fig. 3.10. The adiabatic potential of an electronic state triply degenerate with a doubly
degenerate mode ε_g.

that the vibronic groundstate is a triplet $|T\rangle$) appears to be quite general.
In fact, in the opposite case [$b = 0$ in (3.29)], where no such simple visual-
ization is feasible, the numerical calculations of Englman et al. [43] give
the same result, as does the work of O'Brien [44] for equal coupling to ε_g
and τ_{2g} modes, and Bersuker and Vakhter's perturbative approach [45]. This
is in agreement with our general statement that the Jahn–Teller Hamiltonian
cannot lift the degeneracy. In this regard, it may be interesting to examine
Englman et al.'s result [43] a little closer and from a different point of view.
In their case ($b = 0$, $T \times \tau_{2g}$ problem), we know from the static problem that
there are *four* equi-energetic minima of energy; this means that there are
four different positions in space where the system has its lowest energy or,
in other words, that the groundstate is fourfold degenerate [as long as the
kinetic energy of the nuclei is neglected (static problem)]. Since O_h has only
threefold irreducible representations we are led to the conclusion that for
$b = 0$ (or in general for $b^2/K_\varepsilon > c^2/K_\tau$) the Hamiltonian (3.29) has symmetry
higher than O_h. The fact that such higher symmetry cannot be observed is
due to two circumstances: (a) if dynamical effects are important, tunnelling

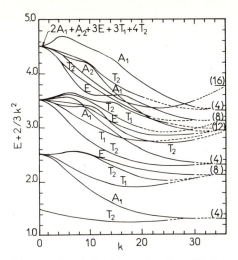

Fig. 3.11. Behavior of lowest vibronic levels as a function of J–T coupling strength for the $T \otimes \tau_{2g}$ problem.

among the four minima splits the quarter into a triplet and a singlet (see fig. 3.11); in group theoretical parlance this means that the kinetic Hamiltonian has the true O_h symmetry; (b) if the kinetic Hamiltonian is really negligible and the effect static, we will observe the system in *one* of the four distorted configurations because of "state preparation": in fact the choice of the particular minimum will depend on a series of causes that we can consider as "initial conditions", which have lower symmetry with respect to O_h. Thus, both in (a) and (b) the whole Hamiltonian (including state preparation) does not have the higher symmetry proper to H_{JT}. As a final remark before we pass on to more practical questions, we note that if tunnelling between different minima is also possible then $b^2/K_\varepsilon > c^2/K_\tau$, i.e. when there are *three* tetragonal minima, but in this case of course, tunnelling must not result in the splitting of the vibronic triplet: this is by no means formally obvious from the point of view of tunnel effect theory and is apparently attained by the special boundary conditions one has to impose on the problem [46]. Now consider a system with weak enough JTE so that dynamical effects are essential; and let us ask ourselves whether or not the JTE can have observable effects: in fact, as we have seen in the previous discussion, the symmetry is preserved in this case during the timescale of a measurement. The answer is positive, and the corresponding theory is probably the major attainment of the last decades in the field of J–T studies of isolated systems [32].

For simplicity let us assume the case of $B^2/K_\varepsilon > c^2/K_\tau$, so that the ground-state is of the type (3.30), and consider an *electronic* operator, for instance the angular momentum, which has only off-diagonal matrix elements

$$\langle \mu | L_v | \lambda \rangle = i\hbar \, \varepsilon_{\mu v \lambda} \, ; (\mu, v, \lambda) \equiv (x, y, z). \tag{3.31}$$

Now, if the system is actually in the state (3.30), the matrix elements of L between any two components of (3.30) are e.g.

$$\langle T_x | \mathbf{L} | T_y \rangle = \langle x | L_z | y \rangle \langle u_1 | u_2 \rangle \langle \varphi_0 | \varphi_0 \rangle. \tag{3.32}$$

The first and third terms of the product are harmless; $\langle x | L_z | y \rangle$ is known and $\langle \varphi_0 | \varphi_0 \rangle = 1$; but due to the selective displacement of the respective minima $\langle u_1 | u_2 \rangle < 1$ because the vibrational wavefunctions only partially overlap. Since the displacement of the minima is the result of the (static) JTE, we see that even in the dynamic regime the effect manifests itself by the quenching of off-diagonal matrix elements of electronic operators. In this simple case the reduction factor $\langle u_1 | u_2 \rangle = K(T)$, which is the same for any electronic operator transforming like the irreducible representations T_1 or T_2 of O_h, can be calculated exactly for the vibronic groundstate and turns out to be [32]:

$$K(T) = \exp\{-3E_{JT}/2\hbar\omega\} \text{ (for } T \otimes \varepsilon_g), \tag{3.32}$$

where $\hbar\omega$ is the vibrational quantum for ε_g modes. In the case of very strong JTE $K(T) \to 0$. If the system is at finite temperature $K(T)$ increases because of the larger overlap of excited vibrational states, but in any case there is a reduction. A qualitatively similar behavior is also found in the more complicated $T \otimes (\varepsilon_g + \tau_{2g})$ case, but no exact formula for the reduction factor can be obtained: approximate expressions for $K(E)$, $K(T_1)$ and $K(T_2)$ for $T \otimes \tau_{2g}$ were derived by Ham [32] and read:

$$\begin{aligned} K(E) &\sim K(T_1) \sim \exp(-qE_{JT}/4\hbar\omega), \\ K(T_2) &\sim \tfrac{1}{3}[2 + \exp(-qE_{JT}/4\hbar\omega)]. \end{aligned} \tag{3.33}$$

An extensive experimental verification of Ham's theory for $T \otimes \varepsilon_g$ was given by Sturge [47] who fitted very well the spin–orbit splitting and the intensities and splittings of the Zeeman lines of V^{2+} in $KMgF_3$, the excited $^4T_{2g}$ electronic level of which interacts predominantly with ε_g modes. Besides the angular momentum operator L, other quantities which suffer Ham-quenching are, for instance, external stress and the hyperfine interaction, but we cannot enter into a detailed description of such effects here.

3.4.3. Jahn–Teller effect in an orbital doublet, in a cubic symmetry

When the electronic state is a doublet E, the only J–T active modes are the

ε_g, because

$$[E_g]^2 = \alpha_{1g} + \varepsilon_g$$

and it is easy to see that in the basis of the electronic doublet $|E_\varepsilon\rangle, |E_\theta\rangle$ the J–T Hamiltonian is given by:

$$H_{JT} = \tfrac{1}{2}d \begin{vmatrix} -q_3 & q_2 \\ q_2 & q_3 \end{vmatrix}, \tag{3.34}$$

where

$$d = -\langle E_\theta | \partial V/\partial q_3 | E_\theta \rangle + \langle E_\varepsilon | \partial V/\partial q_3 | E_\varepsilon \rangle$$

represents the energy separation of the electronic states for unit distortion. Adding to (3.34) the elastic term $\tfrac{1}{2}K(q_2^2 + q_3^2)$ it is a very straightforward matter to solve the static part of the problem (see e.g. Englman [34]) and after introduction of polar coordinates in the $q_2 q_3$ plane

$$q_2 = \rho \sin\theta; \; q_3 = \rho \cos\theta,$$

the energy surface turns out to be:

$$V^\pm = \pm\tfrac{1}{2}d\rho + \tfrac{1}{2}K\rho^2, \tag{3.35}$$

which has the well-known "mexican hat" shape shown in fig. 3.12. The derivation of the surfaces (3.35) is discontinuous in the origin; the surface V has a continuum of minima on the circle of radius $\rho_0 = d/2K$, but if anharmonic terms in the restoring energy and/or quadratic J–T terms are included one finds that only three minima appear on the circle, with the same tetra-

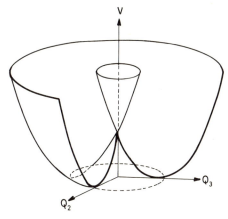

Fig. 3.12. The adiabatic potential of an electronic state doubly degenerate E_g with a doubly degenerate mode ε_g (after Sturge [47]).

gonal symmetry as those of the $T \times \varepsilon_g$ problem. The dynamical problem is not so easily solved: in general, an analytic solution is not possible even in the linear case. Approximate solutions can be found either in the strong or in the weak coupling limit [31, 48, 49, 50]. As for linear coupling, Longuet-Higgins et al. [31] carried out an extensive numerical study of the vibronic energy levels; from their results it appears that the vibronic ground-state is a doublet $|E\rangle$: however as the coupling constant d is increased the energy separation between the groundstate and the nearest excited ones decreases and finally, for $d/\hbar\omega \rightarrow 0$, the groundstate becomes infinitely degenerate. This is the quantum-mechanical manifestation of the existence of an infinity of minima lying on the circle $\rho_0 = d/2K$ that we have mentioned above.

A similar effect arises for non-linear coupling, but in this case the vibronic degeneracy of the groundstate for very strong coupling is of course only threefold. The situation is very analogous to that of strong $T \otimes \tau_{2g}$ coupling. We shall not insist on the $E \otimes \varepsilon_{2g}$ problem, especially in view of the very exhaustive discussion of this case presented by Englman in his book [32]. In the next section we shall rather discuss some of the optical evidence for the JTE.

3.4.4. Optical evidences of the Jahn–Teller effect

As mentioned in § 3.4.2, the Ham-quenching of electronic operators such as the spin–orbit coupling is probably the most general consequence of the dynamical JTE and has been tested experimentally on a number of systems, one of earlier and more extensive works in this sense being Sturge's optical study of $KMgF_3 \cdot V^{2+}$ under an external magnetic field [47]. The JTE however has also more direct consequences on the optical band shapes of impurities in solids; in fact, when one of the states of interest in the optical transition is degenerate the band may be structured and present as many peaks as is the degree of electronic degeneracy. Although this might appear as a trivial consequence of the splitting of the electronic levels operated by the JTE, the situation is not that simple because in some well specified types of coupling (more specifically $T \otimes \varepsilon_g$ and $E \otimes \beta$) the effect leads to no structure of the bands even though the electronic levels for distorted configurations are split.

We shall examine the typical cases $E \otimes \varepsilon_g$ and $E \otimes \beta$ in some detail; this will be done using the semiclassical Condon approximation, as well as the adiabatic approximation: the validity of the latter is certainly limited as regards quantitative predictions, but it is beyond doubt that it leads to qualitatively correct results, providing one is not interested in the vibrational structure, of course, because the nuclear motion is treated classically. The great advantage of the semiclassical methods is that their use makes it possible for non-specialist to understand what is happening.

We consider first the case of a molecule or impurity ion a possessing an electronic state E, at a site of tetragonal symmetry C_4. In such a case, the state E only interacts with a totally symmetrical mode A_1, which is not important, and with a mode of symmetry β: the latter removed the electronic degeneracy via the JTE and the corresponding electron–vibration interaction Hamiltonian is easily written as [33]:

$$H_{el} = \begin{vmatrix} aq & bq \\ bq & -aq \end{vmatrix} + \tfrac{1}{2}Kq^2, \tag{3.36}$$

q being the β-coordinate. This is easily diagonalized to give:

$$V^{\pm} = \pm(a^2 + b^2)^{1/2} q + \tfrac{1}{2}Kq^2. \tag{3.37}$$

Consider now the optical transition from a groundstate of A-type, non-degenerate, to the E-state coupled to the β-mode; the potential surfaces are of the type shown in the fig. 3.13. It is not even necessary to carry on the

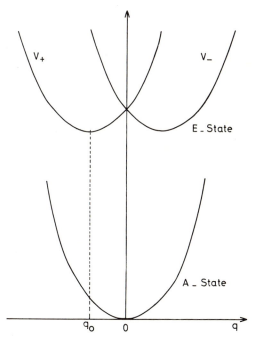

Fig. 3.13. The adiabatic potential of a molecule or an impurity ion possessing an electronic state E_g at a site of tetragonal symmetry C_4 and coupled with a mode of β symmetry.

calculation of the band-shape to understand that in such a transition the structure of the band will not appear. In fact, although for each particular configuration (e.g. q_0) the excited state is split, when the system oscillates in the groundstate it averages, in a sense, the energies of the two split-excited states, which are identical: the result is a single, bell-shaped absorption band. This result, that we have presented on grounds of intuition, can be derived quite rigorously. Let us turn now to the $E \otimes \varepsilon$ case, where the electronic doublet interacts with a doubly degenerate mode.

The interaction Hamiltonian is given by eq. (3.34), and its diagonalization is given by eq. (3.35), which, in terms of the original normal coordinates q_2, q_3, reads:

$$V^{\pm} = \pm \tfrac{1}{2} d(q_2^2 + q_3^2)^{1/2} + \tfrac{1}{2} K(q_2^2 + q_3^2) \tag{3.38}$$

The semiclassical expression for the lineshape is given by [33, 34]:

$$I(\omega) = \sum_f \int \ldots \int dq_1 \ldots dq_k \, P_i(\{q\}) \delta [\hbar\omega - E_f(\{q\}) + E_i(\{q\})]. \tag{3.39}$$

In (3.39), $P_i(\{q\})$ is the probability distribution of the vibrational states of the initial (i) electronic state, $\sum_f$ means sum over all final (f) electronic states, $\{q\}$ represents the entire set of normal coordinates; the δ-factor accounts for energy conservation: $\hbar\omega$ is the light energy, while E_i and E_f are the energies of initial and final states respectively. Without any qualitative change on the results, we take P_i to be a Boltzmann distribution, and insert (3.38) into (3.39):

$$I(\omega) \propto \sum_{\pm} \int\int dq_2 dq_3 (\pi kT)^{-1} \exp[-(q_2^2 + q_3^2)/kT] \, \delta[\hbar\omega - V_{\pm}(q_2, q_3)]. \tag{3.40}$$

Here the factor $(1/\pi kT)$ comes out from the normalization of the Boltzmann distribution, k being the Boltzmann constant. Eq. (3.40) can be integrated exactly [33] yielding

$$I(\omega) = \frac{|\hbar\omega|}{\tfrac{1}{4}d^2 kT} \exp\left(-\frac{(\hbar\omega)^2}{\tfrac{1}{4}d^2 kT}\right), \tag{3.41}$$

which, when plotted against $\hbar\omega$ has a split shape as shown in fig. (3.14). It may perhaps be confusing that two similar potential surfaces as $E \otimes \beta$ and $E \otimes \varepsilon$ give such drastically different results. Actually, the potential for $E \otimes \beta$ looks identical with the cross section of the potential of $E \otimes \varepsilon$ along a plane perpendicular to the $q_2 q_3$ plane. The important difference is that in the $E \otimes \varepsilon$ case one solution of the Schrödinger equation, V^+, for every value of

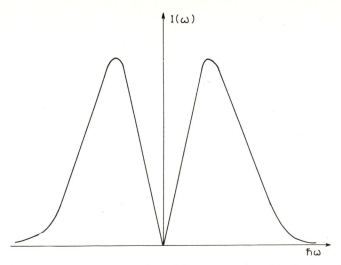

Fig. 3.14. Analytical shape of absorption band for an $A \to E$ transition in the semiclassical theory of Toyozawa and Inoue [33].

$\{q\}$ is higher than the other solution, V^-, contrary to what happens in the $E \otimes \beta$ case, where for positive q, $V^+ \; V^-$ and for negative q, $V^- \; V^+$. In the $E \otimes \varepsilon$ case, the averaging of the excited state energies due to thermal vibrations in the groundstate does not wash away the splitting, because $\langle V^+ \rangle > \langle V^- \rangle$. The difference between the two cases might appear to be evanescent, but as a matter of fact it is the mechanism that causes the band splitting. The band shape for $E \otimes \varepsilon$ has been calculated numerically with a full quantum-mechanical procedure by Longuet-Higgins et al. [31], who found that the doublet structure is just the envelope of the vibronic lines, as shown in fig. (3.15). The classical approximation may be adequate for high temperatures and strong electron–phonon coupling, where the vibrational structure is lost in any case. One significant feature of the quantum-mechanical band-shape, which cannot be reproduced by the semiclassical calculation (at least in the linear J–T approximation) is its asymmetry. From a qualitative point of view, $T \otimes \varepsilon$ behaves like $E \otimes \beta$ (no structure) and $T \otimes \tau$ like $E \otimes \varepsilon$ (triplet structure). The composite $T \otimes (\alpha + \varepsilon + \tau)$ problem has been studied in great detail by Cho [51] who found that the effect of α and ε modes is to smooth the structure due to the τ modes. When the coupling to α and ε is much stronger than to τ the structure tends to disappear. We shall consider these bands again in the next chapter (Tl impurity in alkalihalides). For the moment we just remark that the mechanism leading to the $T \otimes \tau$ triplet is the same as that which we have examined before for $E \otimes \varepsilon$ case.

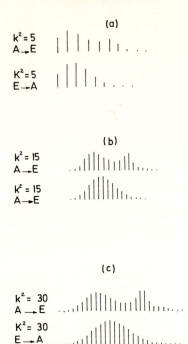

Fig. 3.15. Shape of absorption bands corresponding to a $A \rightleftharpoons E$ transition in cubic symmetry for several coupling with the mode ε_g. (after Longuet-Higgins et al. [31]).

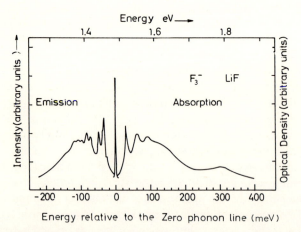

Fig. 3.16. Absorption and emission spectra of F_3^--centers in LiF at 4.2 K. (after Fetterman [26]).

Further evidence for the J–T effect in emission can be found in the breaking down of the "mirror" symmetry between absorption and emission. This effect has been observed for instance in F-center aggregate centers in alkali-halide crystals. These defects are in fact, good systems for the study of J–T effects. They are sufficiently weakly coupled to totally symmetric modes to allow the observation of structures and at the same time have enough coupling to J–T active modes to yield appreciable effects. The F_3^--center studied by the Cornell group [26, 52] is one of the best examples (fig. 3.16). The breaking of mirror symmetry due to the J–T effect is also confirmed by magneto-optical data which show Ham-reduction of orbital g-factor. Another interesting system is the F_3-center [53]. In this case it is the electronic groundstate that is doubly degenerate. Thus it is possible to lift the vibronic degeneracy with uniaxial stress: this stress induces a temperature dependent linear dichroism in the $E \rightarrow A$ absorption band. Since the dichroism depends on the symmetry of the vibrational modes excited, it will vary along the vibrational sideband; this has allowed the identification of the vibrational mode coupled to the electronic groundstate.

The J–T effect is also active in the F-center. Given the O_h symmetry of the excited state (p-like), the J–T coupling can induce structures in the F-band for strong spin–orbit interaction only. This is effectively the case for the F-center in CsF and CsCl. The F^+-center in CaO is also interesting. The spin–orbit interaction is weak, and thus no J–T splitting of the band is possible. At the same time the phonon coupling is also weak and thus both zero-phonon and vibrational sideband structures are observed. It is thus possible to study in detail the effect of external perturbation on the vibronic system and therefore to study quantitatively the Ham effect [54] by measuring the quenching of the appropriate orbital operators. The principle of the measurement is simple since the optical transition to the $^2T_{1u}$ excited state may proceed through the zero-phonon line or through its vibrational sideband. In the first case, the electron arrives directly at the relaxed excited state, i.e. to the ground vibronic level of the excited state, which will have been influenced by J–T effect, including the possible Ham quenching. In the second case we study the non-distorted situation instead, i.e. the non-quenched coupling coefficient [56]. The ratio between the two sets of measurements yields then the appropriate Ham reduction factors. Merle d'Aubigne and Roussel [55] have studied the F^+-band in CaO by analyzing its magnetic circular dichroism at high resolution and using a double resonance technique (light and microwaves, see following chapter). They have found that the spin–orbit coupling in the zero-phonon line is 0.58 cm^{-1} whereas it is -31 cm^{-1} in the sideband: this corresponds to a Ham orbital quenching of $k(T_1) = 0.02$. This quenching is of the same order of magnitude as that observed for instance in the F_3-center [52].

3.5. Resonant Raman scattering

3.5.1. Effect of resonance on the defect-induced Raman scattering

Defect-induced Raman scattering has been discussed in chapter 2. We have shown that defects break both translation and inversion (where it exists) symmetries and lead to a first-order Raman spectrum which reflects to some extent the phonon density of states of the crystal. The Raman event depends on the modulation of the crystal polarizability by vibrations through the transition dipole moments: thus the process is directly related to a breakdown of the Condon approximation. The spectra do not yield information on the electronic states of the system since the polarizability tensor is connected with sums over all the electronic states of the crystal. The formulas discussed in chapter 2 concerning the Raman cross-section are valid only off-resonance, i.e. when the incoming light energy is far from either electronic or vibrational transition energies. If on the other hand the light frequency approaches resonance with one of the electronic transition frequencies ω_{mn}, say $\tilde{\omega}$, it is evident that the term in the polarizability tensor sum within the denominator will grow and eventually dominate the others. At the same time the scattered intensity will increase. This resonance effect is analogous to the resonances in nuclear scattering cross-sections [58] and has important consequences, since it allows the detection of scattering from defects at low concentrations or the selection of the scattering from one defect in presence of other defects for which the resonance condition is not satisfied [59]. There are however other resonance-induced modifications in the scattering which are deeper and more interesting from a fundamental point of view. The theory of resonant Raman scattering is a subject that is too vast for this short section so we shall not dwell on its general aspects here. Rather, we shall briefly discuss a very simple situation which, however, well exemplifies some of the phenomena which take place at resonance. Later we shall slightly generalize to the case of resonant Raman scattering (RRS) from F-centers. Let then $|m\rangle$ be the initial electronic state and $|\bar{n}\rangle$ the resonant state; assuming that all other states are sufficiently removed, we may consider only the leading term in the polarizability, which we may write

$$P_{ij}(\omega_0, q) = \frac{1}{\hbar} \frac{\langle m | M_j | \bar{n} \rangle \langle \bar{n} | M_i | m \rangle}{\tilde{\omega} - \omega_0 - i\Gamma}. \tag{3.42}$$

It is clear that now the scattering will depend on the nature of the electronic state $|\bar{n}\rangle$, and that RRS will give information on this state and its coupling to vibrations. In eq. (3.42), q represents the vibrational coordinates, and

$P_{ij}(\omega_0, q)$ depends on q both through the q dependence of M_i (and this is a deviation from the Condon approximation) and through the q dependence of $\tilde{\omega}(q)$. The latter dependence is generally ignored off-resonance due to the small contribution of each electronic state to the total polarizability. However in the resonance region a small variation in $\tilde{\omega}$ may induce a large variation in the denominator which in turn dominates the polarizability. In fact, for non-degenerate states and reasonably strong coupling this is probably the modulation mechanism which dominates at resonance [60, 61]; thus we may return to the Condon approximation neglecting the q dependence of M_i and write:

$$P_{ij}(\omega_0, q) = C_{ij}[\tilde{\omega}(q) - \omega_0 - i\Gamma]^{-1}, \qquad (3.43)$$

where C_{ij} is essentially a constant which contains the products of the dipole matrix elements associated with the transition $m \rightarrow \bar{n}$. The modulation of P and thus the scattering cross-section will be directly related to the function $\tilde{\omega}(q)$ which in turn is connected with the coupling of $|\bar{n}\rangle$ to totally symmetric vibrations (let us recall that $|\bar{n}\rangle$ is assumed to be non-degenerate). For the RRS, intensity will be directly related to the strength of the electron–vibration coupling for the totally symmetric vibrations which contribute to the spectrum.

Among the relatively few works on RRS from non-degenerate states we may cite the work on RRS from NO_2^- in alkali-halides [62] and that on the F_2^+-center in LiF [58]. In the latter case, the resonance is with the first excited state B_{1u} of F_2^+ using the 6328 line of the He–Ne laser, which is located near the absorption maximum of the multiphonon sideband, thus sufficiently far from the zero-phonon line where interference by luminescence would make the measurement very difficult. The RRS is shown in fig. (3.14) and is independent of light polarization. The spectrum is very similar to that of a one-phonon emission sideband as could be expected since the matrix elements and the selection rules are the same for the two cases.

More interest and more work has been devoted to RRS from degenerate electronic states, and particularly from the first excited state of the F-center in alkali-halides. Remaining in the framework of the simple model discussed before, we note immediately that now non-totally symmetric vibrations will also contribute to the scattering. Thus RRS will yield the electron–vibration coupling strength to all vibrations which couple to the resonant state and broaden its absorption band [59]. Although RRS may be experimentally more difficult, it is the only method to give such coupling directly; stress-induced dichroism in fact will only yield the coupling to long wavelength acoustic vibrations.

The most striking resonance effects are the spectral changes as a function

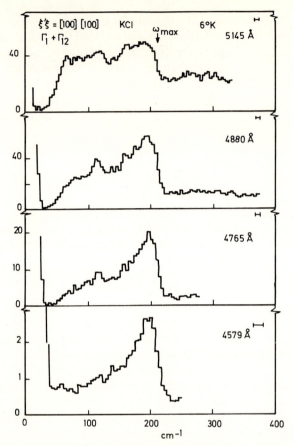

Fig. 3.17. F-center resonant Raman spectra in KCl for several frequencies of excitation (after Buchenauer [63]).

of light frequency in the resonance region. In fig. 3.17 we report the RRS from F-centers in KCl obtained using several lines of the Ar ion laser [63]; the 5154 Å spectrum is very similar to that first reported by Worlock and Porto using the 6328 Å line [64]. Although Buchenauer tried to give a qualitative interpretation of the resonance-induced spectral changes [63] (see also ref. [65]), these results, and the similar ones obtained later by various authors [66], still await a quantitative theoretical explanation. The situation is complicated here by the interplay of spin–orbit and Jahn–Teller interactions in the first excited state and their combined effects on the Raman resonant polarizability. From this point of view a particularly interesting system is the F-center in the cesium halides, where the F-band features strong spin–orbit related structures [67]. This case has been treated

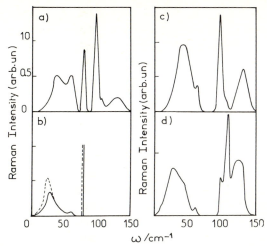

Fig. 3.18. Experimental and calculated Stokes Raman spectra of F centers in CsCl at T = 80 K. (a) 3/4 (T_3^+) experimental spectrum ($\lambda_L = 632.8$ nm). (b) 3/4 (T_3^+) calculated spectra. Full line: $\lambda = -0.67$, $\gamma = -0.05$; dashed line: $\lambda = -0.77$, $\gamma = 0$. (c) (Γ_3^+) experimental spectrum ($\lambda_L = 632.8$ nm). (d) (Γ_3^+) calculated spectrum ($\lambda' = 0$) (after Buisson et al. [69]).

theoretically by Mulazzi and collaborators [67, 68]. Recent experimental results have been reported by Buisson et al. [69], who also offered a semi-quantitative interpretation based on the quasi-resonance approach of Henry [60]. Buisson et al. studied RRS from the F-centers in CsCl and CsBr crystals (and laterly also CsF, J. P. Buisson, private communication), using either an Ar laser or a CW dye laser. The spectra were taken in all the interesting geometries in order to verify the symmetry selection rules off-resonance and their possible breakdown on-resonance. Among other things they found that on-resonance the totally symmetric Γ_1^+ modes (55 cm^{-1} in CsBr and 63 cm^{-1} in CsCl) also gave strong contributions to the forbidden $Z(YZ)Y$ and $Z(YX)Y$ geometries (fig. 3.18). From their data and their theoretical analysis Buisson et al. have concluded that for the cases of CsCl and CsBr, F-centers, the Jahn–Teller effect does not play the dominant role in determining the behavior of the resonance Raman cross-section, which seems to be essentially dominated by the strong spin–orbit interaction in these crystals.

3.5.2. Hot luminescence

A problem faced by anyone who wishes to study RRS is the presence of a sometimes strong background of inelastically scattered light, which must be somehow subtracted from the experimental spectra in order to obtain the "true RRS". Such a background has been termed "hot luminescence" (HL) and its interpretation has been a subject of some controversy. In principle, second-order time-dependent perturbation theory should describe all processes by which the incident phonons interact with the electron–phonon system [61, 65] and thus contain the cross-sections for RRS, hot luminescence and ordinary luminescence. For the simple non-degenerate resonance model we have discussed earlier, it is difficult to make a fundamental distinction between RRS and HL. In fact, optical emission may take place during the vibrational relaxation which follows an optical absorption event: this hot luminescence would be fundamentally similar to RRS for the simple non-degenerate isolated electronic state at resonance with the incoming light. The situation is not so simple and a real distinction probably exists for most other cases (degenerate states, bands etc). A unified theory of RRS and HL has recently been proposed by Toyozawa [70], and we refer the reader to this work for details and for references to previous work on the subject.

Empirically, the various forms of emission are connected with the appropriate characteristic times in the dynamics of the scattering process: the radiative emission time, which is of the order of 10^{-8} s for an allowed transition; the vibrational relaxation time, 10^{-12} s; the Raman time, virtually zero. Thus the type of process will roughly depend on the timescale which we use when looking at it.

From the experimental point of view, HL and RRS have roughly the same intensity and the same spectral location; thus it may be difficult to separate them. Furthermore, they may interfere [71].

Apart from its fundamental interest (the Russian school in particular has worried about this problem, see ref. [62]), HL may yield direct information on vibrational relaxation times and electron–vibration coupling in the excited state. Thus, in spite of the inherent experimental difficulties, a considerable amount of work has been devoted to this subject.

3.5.3. Experimental evidences

In what follows we shall describe some experimental results on HL. In the first F-center RRS investigated by Buchenauer it was already apparent that a non-trivial diffuse inelastic tail existed, the intensity of which grew as resonance was approached. A first explanation for this tail could have been that it was due to multi-quantum replicas of the first-order peaks in the RRS. Such higher order events, and their intensity increase at resonance were already well known in RRS in semiconductors [72, 73]. However, the diffuse tail had a polarization of 0.33, independent of crystal

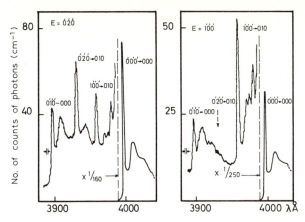

Fig. 3.19. Hot luminescence spectrum of $KCl:NO_2^-$ at 4.2 K for two different excitations. The excitation frequency $E = 1'0'0'$ is smaller than the excitation frequency $E = 0'2'0'$ and then the line $0'2'0' \to 010$ of the hot luminescence spectrum disappears. (after Rebane et al. [74]).

orientation. This led Buchenauer to assume that the tail was due to HL. He then proposed a model in which the scattering mechanism in the excited state was the J–T mixing of 2s and 2p states due to odd LO phonons. Klein and Colwe [73] have observed HL from localized excitons in CdS, by investigating carefully the changes in the lineshape of a particular LO phonon line at resonance. Rebane and collaborators have been studying HL from molecular ion impurities in alkali-halides for many years [74]. In particular the NO_2 ion has a roto–vibrational emission structure which is very well resolved at low temperatures. With very monchromatic light, the ions are excited to a specific vibronic level of the excited electronic state. The level width is very small so that the lifetime of this vibronic level is sufficiently long to allow observation of a radiative transition from this state, yielding a HL zero-phonon line. To this line are then associated one, two etc., quantum lines. These transitions have effectively been observed (fig. 3.19) and their assignment as HL events was based on the following considerations:

(a) the lines shift with the frequency of the exciting light in a manner similar to ordinary Raman scattering;

(b) if the exciting light frequency decreases below that of the appropriate vibronic structure, the HL disappears;

(c) finally, the frequencies of the HL lines are different from those corresponding to the relaxed configuration of the excited state.

From the intensity of the HL lines Rebane and collaborators were able to determine the rotational and vibrational relaxation times (10^{-8} s and 10^{-11} s respectively) of the NO_2^- complex. The long rotational relaxation time indicates, at least in the excited state, that the NO_2^- ion is not coupled to the lattice. This weak coupling is very likely due to the vanishing density of phonon states at the frequency (5 cm^{-1}) of the rotational mode and to the selection rules which govern the roto–vibrational coupling. Since the rotational relaxation time is about the same as the radiative lifetime, it is evident that the emission will take place from a system which is not in thermal equilibrium, which is of course characteristic of HL.

References to chapter 3

[1] F. E. Williams, J. Chem. Phys. **19** (1951) 457; C. C. Klick, Phys. Rev. **85** (1952) 154.
[2] C. Kubo, Phys. Rev. **86** (1952) 929.
[3] Y. Perlin, Sov. Phys. Usp. **6** (1964) 542, and refs. therein.
[4] D. E. Mc Cumber, J. Math. Phys. **5** (1964) 221, 508.
[5] K. K. Rebane, in: Impurity spectra of solids (Plenum Press, New York, 1970).
[6] S. I. Pekar, Sov. Phys. JETP **20** (1950) 519.
[7] K. Huang and A. Rhys, Proc. R. Soc. **A204** (1950) 406.
[8] S. I. Pekar and M. A. Krivoglaz, Trudy, Phys. Instrum. Acad. Sci. Ukr. USSR **4** (1953) 37.
[9] R. H. Silsbee and D. B. Fitchen, Rev. Mod. Phys. **36** (1964) 423.
[10] A. E. Hughes, J. Phys. Chem. Solids **29** (1968) 1461.
[11] Y. Farge, G. Toulouse and M. Lambert, C. R. Acad. Sci. **262** (1966) 1012; D. B. Fitchen, in: Physics of Color Centers, ed. W. B. Fowler (Academic Press, New York, 1968).
[12] Y. Farge, G. Toulouse and M. Lambert, J. Phys. **27** (1966) 287.
[13] G. Dolling, H. G. Smith, R. M. Nicklow, R. R. Vijayaraghava and M. K. Wilkinson, Phys. Rev. **168** (1968) 970.
[14] Y. Farge and M. P. Fontana, Solid St. Commun. **10** (1972) 333.
[15] R. Loudon, Proc. Phys. Soc. **84** (1964) 379.
[16] A. E. Hughes, Ph.D. Thesis, Oxford University (unpublished).
[17] K. K. Rebane (op. cit.) p. 89.
[18] T. Timusk and W. Staude, Phys. Rev. Letters **13** (1964) 373.
[19] A. E. Evans and D. B. Fitchen, Phys. Rev. **B2** (1970) 1074.
[20] V. Narayanamurti, W. D. Seward and R. O. Pohl, Phys. Rev. **148** (1966) 481; W. D. Seward and V. Narayanamurti, Phys. Rev. **148** (1966) 463.
[21] B. Wedding and M. V. Klein, Phys. Rev. **177** (1969) 1274; M. V. Klein, B. Wedding and M. A. Levine, Phys. Rev. **180** (1969) 902.
[22] R. Avermaa and L. Rebane, Phys. Stat. Solidi **35** (1969) 107.
[23] A. A. Kaplyanskii, Opt. Spectrosc. **16** (1964) 329.
[24] D. B. Fitchen, in: Physics of Color Centers (op. cit.).
[25] M. Maki, Y. Farge and M. P. Fontana, Int. Conf. Color Centers Reading (1971).
[26] H. R. Fetterman, Thesis, Cornell University (1968).
[27] A. W. Overhauser and H. Ruchardt, Phys. Rev. **112** (1958) 722; A. A. Kaplyanskii and V. N. Medvedev, Sov. Phys. Sol. St. **11** (1969) 109.
[28] H. A. Jahn and E. Teller, Proc. R. Soc. **A161** (1937) 220.
[29] H. A. Jahn, Proc. R. Soc. **A164** (1938) 117.
[30] W. Moffit and A. D. Liehr, Phys. Rev. **106** (1957) 1195.
[31] H. C. Longuet-Higgins, V. Opik, M. H. L. Pryce and R. A. Sack, Proc. R. Soc. **A244** (1958) 1.
[32] F. S. Ham, Phys. Rev. **138A** (1965) 1727.
[33] Y. Toyozawa and M. Inoue, J. Phys. Soc. Japan **20** (1965) 1289; **21** (1966) 1663.
[34] R. Englman, in: The Jahn–Teller Effect in Molecules and Crystals (Interscience, New York, 1972).

[35] J. H. Van Vleck, J. Chem. Phys. **7** (1939) 72.
[36] V. Opik and M. H. L. Pryce, Proc. R. Soc. **A238** (1957) 425.
[37] A. Ranfagni and G. Viliani, Phys. Rev. **B9** (1974) 4448.
[38] P. Wysking and K. A. Muller, Phys. Rev. **173** (1968) 327.
[39] I. B. Bersuker and V. Z. Polinger, Phys. Letters **A44** (1973) 495.
[40] M. Bacci, A. Ranfagni, M. P. Fontana and G. Viliani, Phys. Rev. **B11** (1975) 3052.
[41] M. Bacci, B. D. Bhattachakyya, A. Ranfagni and G. Viliani, Phys. Letters **55A** (1976) 489.
[42] A. Ranfagni, Phys. Rev. Letters **28** (1972) 743.
[43] R. Englman, M. Carrer and S. Taoff, J. Phys. Soc. Japan **29** (1970) 306.
[44] M. C. M. O'Brien, Phys. Rev. **187** (1969) 407.
[45] I. B. Bersuker and B. G. Vekhter, Phys. Stat. Solidi **16** (1966) 63.
[46] M. C. M. O'Brien, (1977) private communication.
[47] M. D. Sturge, Phys. Rev. **B1** (1970) 1005.
[48] W. Moffitt and W. Thorson, Phys. Rev. **108** (1957) 1205.
[49] M. C. M. O'Brien, Proc. R. Soc. **A281** (1964) 323.
[50] M. Wagner, Phys. Letters **29A** (1969) 472.
[51] K. Cho, J. Phys. Soc. Japan **25** (1968) 1372.
[52] J. A. Davis, Thesis, Cornell University (unpublished).
[53] R. H. Silsbee, Phys. Rev. **138A** (1965) 180.
[54] F. S. Ham, Phys. Rev. **138A** (1965) 1727.
[55] Y. Merle d'Aubigne and A. Roussel, Phys. Rev. **B3** (1971) 1421.
[56] C. H. Henry, S. E. Schnatterly and C. Slichter, Phys. Rev. **137A** (1965) 583.
[57] J. M. Blatt and W. F. Weisskoff, Theorectical Nuclear Physics (J. Wiley, New York, 1952) p. 394.
[58] Y. Farge and M. P. Fontana, Solid St. Commun. **10** (1972) 333.
[59] C. H. Henry, Phys. Rev. **152** (1966) 699.
[60] V. Hizhnyakov and I. Tehver, Phys. Stat. Solidi **21** (1967) 755 and ref. therein.
[61] K. K. Rebane, Impurity Spectra of Solids (Plenum Press, New York, 1970).
[62] C. J. Buchenauer, Thesis, Cornell University (1970).
[63] J. M. Worlock and S. P. S. Porto, Phys. Rev. Letters **15** (1965) 697.
[64] V. Hizhnyakov and I. Tehver, in: Light Scattering in Solids, ed. M. Balkanski, (Flammarion, Paris, 1971) p. 57.
[65] J. P. Buisson, A. Sadoc, L. Taurel and M. Billardon, in: Light Scattering in Solids, ed. M. Balkanski, R. C. C. Leite and S. P. S. Porto (Flammarion, Paris, 1975) p. 587; D. S. Pan and F. Lüty, ibid p. 540; J. P. Buisson, S. Lefrant, A. Sadoc, L. Taurel and M. Billardon, Phys. Stat. Solidi (b)**78** (1976) 779.
[66] H. Rabin and J. H. Schulman, Phys. Rev. **120** (1960) 2007; P. R. Moran, Phys. Rev. **137A** (1965) 1016; D. B. Fitchen, Phys. Rev. **133A** (1964) 1599; K. Cho, J. Phys. Soc. Japan 25 (1968) 1372.
[67] E. Mulazzi and M. F. Bishop, J. Phys. C: Solid St. Phys. 7 (1976) 1; E. Mulazzi and N. Terzi, Solid St. Commun **18** (1976) 721.
[68] J. P. Buisson, S. Lefrant, M. Ghomi, L. Taurel, and J. Chapelle, Proceedings. Int. Conf. Lattice Dynamics, Paris (1977), Flammarion Sciences (1978) 223. S. Lefrant, J. P. Buisson and L. Taurel, C. R. Acad. Sci. **B282** (1976) 407.
[69] Y. Toyozawa, J. Phys. Soc. Japan **41** (1976) 400.
[70] K. K. Rebane, in: Light Scattering in Solids, ed. G. Wright (New York, 1971) p. 513.
[71] J. F. Scott, R. C. C. Leite and T. C. Damen, Phys. Rev. **188** (1969) 1285.
[72] M. V. Klein and J. P. Colwell, in: Light Scattering in Solids, ed, M. Balkanski (Flammarion, Paris, 1971) p. 72 and refs. therein.
[73] P. M. Saari and K . K. Rebane, Solid St. Commun 7 (1969) 887.
[74] K. K. Rebane, R. A. Averma, L. A. Rebane and P. M. Saari, in: Light Scattering in Solids, ed. M. Balkanski (Flammarion, Paris, 1971) p. 72; K. K. Rebane, in: Luminescence of Inorganic Solids. (Plenum Press, N.Y., 1978).

4 THE ELECTRONIC STATES OF DEFECTS IN IONIC SOLIDS

4.1. Introduction

A point defect in a crystal will modify its electronic properties both statically and dynamically. Statically, the defect will destroy its translation symmetry and also where it exists, its inversion symmetry. Dynamically, the point defect may be coupled to the surrounding ions by forces that are different from those coupling the ion it replaces. These effects yield the localized and resonant vibrational modes we have discussed in chapter 2. In this chapter we shall discuss the corresponding effects for the electronic states of the host crystal.

In an ionic insulator, the size of the forbidden band gap is always of the order of several eV, going to a maximum of about 10 eV for LiF. Thus it is likely that gap modes will be caused by the defect. Resonant states in the conduction band are also possible, in conjunction or not with gap modes. Another classification of defect-related electronic states may consist in whether or not the defect has its own electronic degrees of freedom. The case of internal degrees of freedom, analogous to the case of molecular vibrations, is associated with atomic or molecular impurities. The other case is that of color centers. In this chapter we shall treat in some detail an important example from each of the two categories, trying thus to give a general idea of the type of physics involved. Again, we shall considerably restrict our analysis so that, instead of superficially discussing the electronic states of all the defect centers known, we shall treat in some depth some well defined, but important subjects.

The discussion shall begin with the F-center in the alkali-halides; this defect is the prototype color center and in many of its characteristics one of the most interesting. This center yields both gap and resonant modes (known as F- and L-bands respectively). As the typical defect with internal degrees

of freedom, we have chosen the Tl impurity in alkali-halides, studied inten-
sively for at least forty years, and understood quantitatively only recently.
We shall emphasize the available experimental information and the models
developed to interpret it. For calculations and more detailed theories,
the reader is referred to the literature [1–4]. At the end of this chapter, we
discuss two important problems. The first one is the self-trapped excitons in
alkali-halides; the excitonic recombination can give rise to the well known
photochemical process (formation of a pair F + H) and then it is necessary
to understand as well as possible the electronic structure of the exciton.
The second one is the non-radiative transition problem; this very important
phenomenon will be discussed through the triplet to singlet transitions of
F_2-centers in alkali-halides.

4.2. The F-center in alkali-halides

We shall discuss this subject, mixing on purpose models and experiments,
or more exactly we shall go back and forth between theoretical models and
experimental evidence; in fact, the simplest models for the F-center have
generated several experiments whose results were not completely inter-
pretable with these models. So more elaborate models were devised, which
then were tested by further more sophisticated experiments.

4.2.1. The simple models

Among the simple models for the F-center, one of the simplest is the continuum model. In this model, the F-electron wavefunction is assumed to be
sufficiently delocalized so that the effective mass approximation can be
made and furthermore the crystal behaves relative to the electron as a
dielectric continuum of defined dielectric constant. The anion vacancy
will then appear as a positive point charge and the delocalized electron
will be bound to the vacancy by the modified Coulombic interaction. Thus
the problem is reduced to that of the hydrogen atom and the relative energy
levels will be distributed according to:

$$E_n = -(e^4 m^*/2\varepsilon_\infty \hbar^2)(1/n^2). \tag{4.1}$$

The F-absorption band corresponds to the 1s–2p transition and the transition energy turns out to be:

$$E_F = -3e^4 m^*/8\varepsilon_\infty^2 \hbar^2, \tag{4.2}$$

The energy of the maximum of the F-band will then be proportional to the inverse square of the dielectric high frequency constant for all the alkali-halides. This prediction is however in contrast with experiment. This negative result is important since it indicates that one of the fundamental hypotheses of the continuum model, i.e. the extended character of the wavefunction, is wrong. The F-electron must be strongly localized inside the vacancy both in its groundstate and in the first excited state. This localization is well known and for the 1s state has been determined directly by ENDOR. Thus a better approximation is the particle in a box, in which the potential is that due to the six nearest neighbors around the vacancy, with a characteristic dimension that is well approximated by the nearest neighbor distance a. In this model the energy spacing between the two lowest levels, i.e. the F-band energy, is given by:

$$E_F = 3\pi^2\hbar^2/8ma^2, \tag{4.3}$$

where m is the real electron mass. The particle in a box model (extreme case of the tight-bonding approximation) predicts then that the E_F band energy will scale in the different alkali-halides as the inverse square of the lattice parameter; experimentally, this energy is found to vary as $a^{-1.88}$ (Mollwo–Yvey law).

The box model is thus in excellent agreement with experiment considering the approximations made. This agreement implies that all the more sophisticated theories will have to use a potential which will reduce to a square well in the vacancy, since the resulting wavefunctions must be localized on the vacancy, at least for the lowest lying states. Of course a more sophisticated theory will also have to account for the finer characteristics of the F-center, such as the nature of the higher excited states, the spin–orbit coupling and the emission properties.

4.2.2. The first excited state; spin–orbit coupling, angular momentum, Stark and Zeeman effects, uniaxial strain effects

The first excited state may be considered as a 2p state in cubic symmetry. Its actual wavefunction being constructed by a suitable linear combination of "p" wavefunctions of the neighboring alkali-ions. In order to determine quantitatively such wavefunctions, it is necessary to study the effect of electric and magnetic fields, since the response of the electronic state is directly related to its orbital momentum eigenvalues. For the F-center, as for the majority of color and impurity centers, the electric or magnetic perturbation energy is several orders of magnitude smaller than the optical bandwidth, this latter one due to the strong electron–vibration coupling. The resulting difficulties are both experimental and theoretical. Experi-

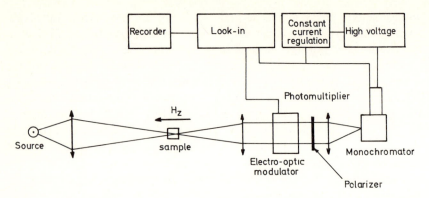

Fig. 4.1. Typical experimental set-up to measure Magnetic Circular Dichroism (MCD). The electro-optic light modulator acts as an alternate $\pm\lambda/4$ plate and, together with the linear polarizer at 45° from its neutral lines, will analyze $\sigma^{\pm}$ circularly polarized light yielding an AC signal proportional to $I_+ - I_-$, which is finally detected by the lock-in amplifier. The photomultiplier voltage is regulated in such a way that the average output photocurrent is kept constant: this allows the direct measurement of $(I_+ - I_-)/(I_+ - I_-)$.

mentally, very small relative changes (sometimes inferior to 10^{-4}) in the optical band-shapes must be detected. Theoretically, such variations must be related to the interesting quantities such as the spin–orbit splitting in the 2p state, the magnitude of g_{orb} and the possible mixing of other electronic states.

The experimental problem was solved by the introduction of the lock-in amplifier [5]. In fig. 4.1, we show the scheme of a set-up to detect the small circular dichroism induced in the F-band by an applied magnetic field; with such a system, polarization degrees as small as 10^{-4} may be measured with good signal to noise ratio.

Theoretically, the use of the method of moments [6] has greatly facilitated the understanding of the effect of applied perturbations on the optical transitions [7]. The basic idea of such a method is to study the effect of the perturbation on the moments of the band instead than on the band itself. The moment evaluation involves sums on the complete set of vibrational states which couple to the electronic state; thus it is not necessary to know the detailed nature of the vibrational wavefunctions. The moment variations are then directly related to the quantities which define the response of the electronic state to a given external perturbation; furthermore it is not necessary to know in detail the electron–vibration interaction which causes the broadening of the optical band. We shall now give some definitions. If $\alpha_\eta(E)$ is the absorption coefficient for light having polarization η and energy E, then the shape function $f_\eta(E)$ is defined by:

$$\alpha_\eta(E) = CEF_\eta(E), \tag{4.4}$$

where C, is approximatively a constant in the visible spectrum. The zero moment of $f_\eta(E)$ is defined by:

$$A_\eta = \int f_\eta(E)\, dE \tag{4.5}$$

and the first and second moments by:

$$\langle E_\eta \rangle = A_\eta^{-1} \int f_\eta(E)\, E\, dE, \tag{4.6}$$

$$\langle E_\eta^2 \rangle = A_\eta^{-1} \int f_\eta(E)\, E^2\, dE. \tag{4.7}$$

The zero moment is thus connected with the absorption intensity, i.e. the transition oscillator strength, the first moment to the position of the maximum of the band and the second moment to the bandwidth.

4.2.2.1. The Zeeman effect.

The problem of the effect of a magnetic field on the 2p-like excited state of the F-center was treated in detail by Henry et al. (HSS) [7], and we shall briefly summarize their results here. Since the magnetic perturbation energy is of the same order of magnitude as the F-center spin–orbit interaction energy, both must be included in the perturbing Hamiltonian H':

$$H' = \mu H_z L_z + \lambda L_z S_z, \tag{4.8}$$

where the field direction defines the z-axis, μ is the Bohr magneton and λ the spin–orbit parameter. The application of the method of moments yields the following results:

there is no zero-moment change if the interaction with higher energy states is neglected.

the first-moment change for circularly polarized light is:

$$\langle \Delta E \rangle_\pm = \pm(g_{\text{orb}}\mu H_z + \tfrac{2}{3}\delta\langle S_z \rangle), \tag{4.9}$$

where δ is the excited state spin–orbit splitting (and not the coupling constant) and $\langle S_z \rangle$ is the groundstate spin polarization

$$\langle S_z \rangle = -\tfrac{1}{2}\tanh(\mu H_z / kT). \tag{4.10}$$

Eq. (4.9) has a simple physical meaning. The first term is just the ordinary diamagnetic Zeeman effect on the 2p state, as for instance in the alkali atoms. The second term, which is paramagnetic, is due to the spin–orbit interaction in the excited state. In fact, the spin–orbit term may be considered as an effective magnetic field:

$$H_{\text{SO}} = (\lambda/\mu) S_z, \tag{4.11}$$

due to each electron spin. Such an internal field will not yield any effect in

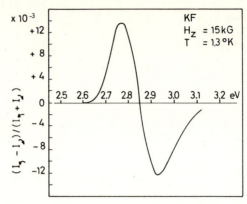

Fig. 4.2. MCD signal due to F-centers in KF [after M. P. Fontana, Phys. Rev. **B2** (1970) 1107)].

zero-external field since the spins will be directed at random. Under an applied field, the spins will tend to align along its direction and thus at sufficiently low temperatures, a net polarization, given by eq. (4.10) will result. This polarization will be associated with a non-zero internal magnetic field which will add to the external field and contribute to the Zeeman effect in the excited state. The dichroism signal resulting from the different absorption for left and right circularly polarized light is shown schematically in fig. 4.2, for the case of the F-center in KF; the equipment used is that described in fig. 4.1. Using eq. (4.9), it is possible to obtain from such measurements the values of g_{orb} and of the spin–orbit splitting by studying the behavior of the dichroism as a function of field and temperature.

These measurements have been made for the F-center in most of the alkali-halides [8] and the most important results are summarized in table 4.1.

It is well known that absorption and dispersion are related by the Kramers–Krönig relations; in the F-band region there must then be a Faraday rotation associated with the circular dichroism. As a matter of fact the first magneto-optical measurements on the F-center consisted in the detection of Faraday rotation induced by the magnetic field [9]. It is important however to realize that the Kramers–Krönig relations are not strictly valid in a magnetic field (such a field destroys time inversion symmetry [5]) and thus it is not immediate in obtaining the circular dichroism from Faraday rotation data and vice versa. This point has been discussed by Smith [10].

From these measurements we see that the spin–orbit splitting is anomalously large and has the opposite sign from that expected on the basis of the alkali-atom model for the F-center; in the F-center, the $P_{3/2}$ component of the spin–orbit doublet is below the $P_{1/2}$ level. The size of the spin–orbit

Table 4.1

g_{orb} and spin–orbit splitting δ of the unrelaxed excited state of F-center in some alkali-halides.

Crystal	δ (Faraday rotation) meV	δ (Circular dichroism) meV	g_{orb}
LiF	-2.5	-3.4[a]	0.59[a]
NaCl	-5.1 ± 1.0	-7.7 ± 1.0	0.38
NaBr	—	-28.3 ± 6	—
KF	—	-3.2[b]	1[b]
KCl	-11.4 ± 2.3	-10.1 ± 1.0	0.62
KBr	-19.2 ± 3.8	-29.8 ± 4	0.47
KI	-30.0 ± 3.2	-57.0 ± 8	0.83
RbBr	-32.4 ± 6.5	-26.6	0.47
RbCl	-15.1	-15.1	—
		-24.1 ± 4	
RbI	-50	—	
CsBr	-42 ± 8.4	-40.9 ± 5	—
CsCl	—	-37.2 ± 5	—

[a]G. A. Osborne, B. D. Bird, P. J. Stephens, J. J. Duffield and A. Abu-Shumays, Solid St. Commun. **9** (1971) 33.
[b]M. P. Fontana, Phys. Rev. **B2** (1970) 1107.
Other values are extracted from the paper of Henry and Slichter (table 6.1) in: Physics of Color Centers (op. cit).

splitting seems to depend strongly on the nature of both the nearest cations and the next nearest anions. These results indicate that the spin–orbit interaction in the first excited state of the F-center is essentially due to the surroundings of the F-center and not to its intrinsic degrees of freedom. Smith [11] has proposed a theory which interprets this result and which leads to good agreement with experiment. The essential point of the theory is the following: the F-center wavefunction must be orthogonalized to the electronic states of the shells of nearest and next nearest neighbors, as a consequence of the exclusion principle. This orthogonalization introduces in the wavefunction terms that lead to currents in the neighborhood of the near neighbor shells; such currents interact with the strong electric fields present in the immediate environment of the nuclei and thus lead to large values of the spin–orbit splitting. The negative sign is now easy to understand: as seen from the ions, the currents move in a direction which is opposite to that observed by the vacancy. In crystals made of heavier elements, the spin–orbit interaction becomes so high ($\delta > 600$ meV in CsCl [12, 13]), that the F-band is structured (actually, it is resolved into three sub-bands and not into a doublet due to the Jahn–Teller effect, see previous chapter).

4.2.2.2. The Stark effect. Since the F-center has inversion symmetry, it may not have a first-order Stark effect on the 2p state [14]. However, if a state of opposite parity was sufficiently close to the 2p level, it would be possible to detect the possible mixing of these two levels by an applied electric field. This mixing would cause an increase in the oscillator strength of the transition to the odd level (the 2p state), and level repulsion between the two levels, which would increase the bandwidth. These effects should cause a linear dichroism in the F-band, and such a dichroism was effectively observed by Chiarotti et al. [15] in KCl. These authors have applied an electric field perpendicularly to the direction of propagation of the incident light. This field induces a small modulation in the transmitted light; such a modulation is detected by means of a lock-in amplifier at a frequency which is double that of the electric field frequency.

The observed variations in the absorption coefficient were of the order of 10^{-4}. A typical result is shown in fig. 4.3, for a crystal of KCl at 55 K in an AC electric field of $40 \, \text{kV} \, \text{cm}^{-1}$, with the electric vector of the incident light parallel to the direction of the applied field. As predicted, the signal increases with the square of the applied field intensity. A moment analysis of the signal in fig. 4.3, shows that neither the zero or the first moments of the F-band vary appreciably and that the dichroism signal is essentially due to a second-moment change as discussed previously and as demonstrated theoretically by HSS. The second-moment variation is due to mixing of the $2p_z$ component of the 2p state with the 2s state nearby; this last level would be above the 2p level by about 0.1 eV. This energy is sufficiently large that, in most cases, the properties of the F-center absorption are determined by the 2p state alone. In any case a great merit of the work of Chiarotti et al. has

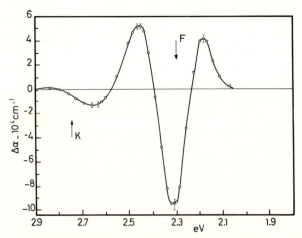

Fig. 4.3. The relative change of the absorption coefficient of the F- and K-bands in KCl due to an external electric field [after G. Chiarotti et al. Phys. Rev. Letters **17** (1966) 1043].

been to show the existence of the 2s state very near in energy to the 2p state. As we shall see later, this information was extremely useful in studying the nature of the relaxed excited state of the F-center.

4.2.2.3. Uniaxial stress effects. One of the most evident characteristics of the F-band is its width (about 0.2 eV in most alkali-halides). The fundamental causes of this width have been discussed in chapter 3. It is possible to determine experimentally the various contributions to the bandwidth due to the vibrations of different symmetry which couple to the F-center electronic state; for that purpose an external stress is applied parallel to the various crystallographic directions, and one then observes the ensuing modifications of the F-band. An applied stress should induce a linear dichroism in the F-band; such dichroism has indeed been observed by Schnatterly [16] on the F- and K-bands in several alkali-halides, and the data were interpreted on the basis of the method of moments. Gebhardt and Meier [17] have also studied this effect; however they interpreted their results by assuming a rigid shift of the F-band under stress. If their results are reinterpreted with the method of moments they turn out to be in good agreement with Schnatterly's. The stress causes a first-moment change in the F-band for linearly polarized light; the change is proportional to the stress intensity. Using this linear variation for stresses in the [100], [110], and [111] directions, it is possible to obtain the values of the coupling coefficients of the excited state with the lattice distortions of the appropriate symmetry, i.e. for a cubic crystal A_{1g}, E_g, T_{2g}. HSS have shown that the ground electronic state is not affected by the applied stress in first order, thus one can neglect it in the calculations. The coupling coefficients for the 2p-like excited state

Table 4.2
The elastic coupling coefficients of the excited state of the F-center in alkali-halides and the lattice vibrational contributions to the second moment of the F-band[a].

	$A_1(10^4 \text{eV kg}^{-1}\text{mm}^{-2})$	$A_3(10^4 \text{eV kg}^{-1}\text{mm}^{-2})$	$A_5(10^4 \text{eV kg}^{-1}\text{mm}^{-2})$
KCl	13.6 ± 0.5	6.1 ± 0.9	8.8 ± 1.2
KBr	14.6 ± 0.5	4.4 ± 0.6	6.3 ± 0.9
KI	17.6 ± 0.5	6.2 ± 0.9	9.8 ± 1.4
NaCl	16.8 ± 0.5	2.4 ± 0.4	6.7
RbCl	13.9 ± 0.5	7.0 ± 1.0	5.8 ± 0.8
	$E_1^2(\text{eV})^2$	$E_3^2(\text{eV})^2$	$E_5^2(\text{eV})^2$
KCl	0.058 ± 0.006	0.032 ± 0.003	0.039 ± 0.004
KBr	0.073 ± 0.007	0.019 ± 0.002	0.029 ± 0.003
KI	0.058 ± 0.006	0.028 ± 0.003	0.026 ± 0.003
NaCl	0.054 ± 0.006	0.029 ± 0.003	0.033 ± 0.003
RbCl	0.10 ± 0.016	0.046 ± 0.006	0.040 ± 0.001

[a]The reader will find the nomenclature in chapter 2 (identical to the parameters b of table 2.4) and in Schnatterly's paper 16 .

determined in this fashion by Schnatterly are shown in table 4.2. In this table, B_1 is the coefficient for coupling to the totally symmetric distortions, B_2 and B_3 the corresponding coefficients for the E_g and T_{2g} distortions. In the same table are reported the contributions of the lattice modes of the three symmetries considered to the second moment of the F-band. The non-symmetric vibrations are seen to contribute to the F-bandwidth at least as much as the totally symmetric vibration. This result was very important in that all the simple "configuration coordinate models" for the F-center optical transitions assumed coupling only to the totally symmetric vibration (the "breathing mode"). Finally, we must recall that in principle the stress coupling coefficients refer only to acoustic vibrations of very large wavelength; the extrapolation of the results to other types of vibrations as for instance LO phonons is an approximation which should be verified in each case.

4.2.3. F-center luminescence

The differences between optical transitions in isolated atoms and defects in solids are particularly evident when considering the characteristics of the F-center luminescence. For instance, if such emission were similar to that of an isolated atom, the emission energy would follow a Mollwo–Yvey law, like the peak absorption energy. Furthermore, since the F-band oscillator strength is almost unity, the decay time of the luminescence should be that of an allowed transition, i.e. approximatively 10^{-8} s. in the energy domain of the F-emission. Unfortunately, there is no correlation between the emission energies and the lattice parameter in the various alkali-halides, and furthermore the measurement by Swank and Brown [18] of the decay time of the F-center emission in KCl yielded the surprisingly low value of about 10^{-6} s.

The radiative lifetime of the emitting state is directly related to its wavefunction by:

$$1/\tau_R \sim |\langle f | M | i \rangle|^2, \tag{4.12}$$

where $|f\rangle$ and $|i\rangle$ are the final and initial state wavefunctions respectively and M is the dipole operator. The large discrepancy between the decay time of the emission from the relaxed excited initial state of the F-center (F*) may be explained by a substantial difference between the excited state in absorption (i.e. before relaxation) and F*. This difference must then take place during the nuclear relaxation in which the electronic state adiabatically follows the nuclear deformation. At the end of the relaxation process, the nuclei will be in their new equilibrium configuration, to which will correspond a wavefunction which may be different both in its spatial and angular dependence from that of the unrelaxed excited state reached in the absorption event. This is certainly the case for the F-center, since the Mollwo–Yvey law is not followed in emission and the excited state lifetime is anomalously long. This problem has stimulated much work devoted to giving at least a semi-quantitative explanation of the nature of the F-state. We shall now recall briefly the main characteristics of the F-center luminescence, before going on to discuss the more recent experiments and theories which have greatly increased our understanding of the relaxed excited state of the F-center in alkali-halides.

The emission, which is characterized by a large Stokes shift is generally located in the near IR (from 0.7μ to 1.3μ). The luminescence efficiency is essentially unity at sufficiently low temperatures [20] (fig. 4.4a) (with the possible exception of NaF); there is therefore an emitted phonon for each phonon absorbed in the F-band. This yield η remains constant up to about 100 K, at which temperature it begins to decrease exponentially with increasing temperature, so that above about 200 K, no more emission is observed. The luminescence decay time also has the same behavior as can be seen in

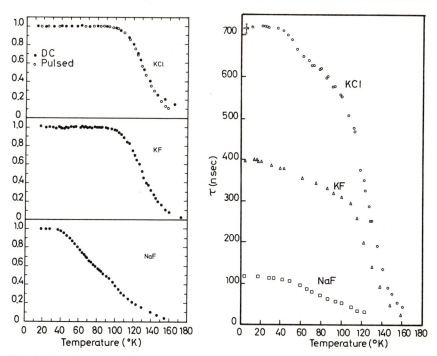

Fig. 4.4. Luminescence characteristics of the F-center in some alkali-halides (after Stiles et al. [22]). (a), T dependence of the yield η; (b), T dependence of the decay time τ.

fig. 4.4b, relative to the F-center in KCl [18, 21, 22, 23]. Neglecting for the moment the small decrease in τ below 90 K, the results shown in fig. 4.4, indicate that the rapid decreases of both η and τ at high temperatures are due to the same mechanism. Swank and Brown have proposed a two-channel decay model to interpret these effects: the first channel is the radiative decay to the groundstate with an associated transition probability $1/\tau_R$ (not $1/\tau$); the second channel corresponds instead to non-radiative transitions which empty the F-state at a rate $1/\tau_{NR}$. Photoconductivity measurements performed together with the photoluminescence ones by Swank and Brown have shown that the non-radiative process is the thermal ionization of the F-center electron into the conduction band. The probability is then:

$$1/\tau_{NR} = A \exp(-U/kT), \tag{4.13}$$

where the activation energy U is the separation energy between the F-level and the conduction band minimum, and A is the characteristic infinite temperature jump frequency. Within this model the total decay probability becomes:

$$1/\tau = (1/\tau_R) + A \exp(-U/kT). \tag{4.14}$$

At very low temperature $(1/\tau) = (1/\tau_R)$ and thus no F*-electron is thermally ejected into the conduction band. Now the quantum yield η may be written as:

$$\eta = \tau/\tau_R \tag{4.15}$$

and thus from (4.14) and (4.15) we obtain:

$$(1 - \eta)/\tau = A \exp(-U/kT). \tag{4.16}$$

If the two-channel model is correct, a semilog plot of $(1 - \eta/\tau)$ versus $1/\tau$ should yield a straight line, from which the activation energy U and the jump frequency A, may be obtained. This is effectively the case for a great number of alkali-halides. The graph is shown in fig. 4.5, for three alkali-halides and the values for U, A and τ_R obtained by several authors are reported in table 4.3.

The two-channel model predicts that both the decay time and quantum yield be constant at sufficiently low temperatures. From fig. 4.4, it is clear that, when the measurements are sufficiently precise, there is a significant variation in τ at temperatures for which η is constant. From the definition of eq (4.15), it is obvious that τ_R is not constant at low temperatures, if we consider as valid the two-channel model.

The F*-wavefunction must then vary with temperature so that the related radiative lifetime will decrease with increasing temperature; alternatively,

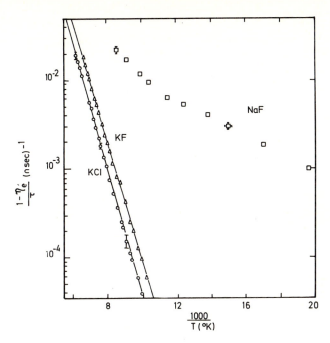

Fig. 4.5. Temperature variation of $(1 - \eta)/\tau$ for the F-center in several alkali-halides. Full line, eq. (4.16), with parameters of table 4.3 (from Stiles et al. [22]).

Table 4.3
Some fundamental parameters of the relaxed excited state of the F-center in some alkali-halides.

	U(meV)	$A (10^{12} s^{-1})$	α^2	δE (meV)
NaF	55[a]	—	0.1[b]	17[b]
KF	131[b]	0.5[b]	0.3[b]	17[b]
				8[e]
KCl	148[b]	1.0[b]	0.5[b]	18[b]
				13[c]
RbCl	130[c]	—	0.3[b]	16[d]

[a] P. Podoni and G. Spinolo, Solid. St. Commun. **4** (1966) 263.
[b] Reference. [23].
[c] G. Spinolo, Phys. Rev. **137A** (1965) 1495.
[d] L. D. Bogan. Thesis (Cornell University, 1968) unpublished.
[e] Y. Ruedin, P. Schnegg, M. A. Aegerter and C. Jaccard, in: Int. Conf. Color Centers in Ionic Crystals (Reading, 1971) abstract 48. They have determined δE from a study of temperature dependence of T_1, the spin–lattice relaxation time, of F^* (see p. 175).

or concurrently the F* state may be made up of two or more closely spaced levels with different characteristic lifetimes. In this last case, the temperature variation of τ_R would be due at least partially to the temperature variation of the relative populations of these levels. Many experiments have been performed to determine the nature of F*; before discussing them, we shall present a very brief outline of the theoretical approaches and resulting models which spurred the more recent experimental work.

4.2.4. Theoretical calculations on the electronic states of the F-center

This section shall be entirely dedicated to the F-center; it is however evident that a great number of the ideas exposed here may be applied to the solution of similar problems in the case of other defects. The calculations on the F-center are numerous and thus we refer the interested reader to the reviews of Gourary and Adrian [2], Markham [3], Fowler [1], Stoneham [4], Opik and Wood [24], Bennett [25] and Petrashen et al. [25]. Here we shall discuss only the fundamental concepts and approximations used in these calculations.

The first question concerning the F-center is: is the F-center electron equivalent to the other electrons in the crystal? If the answer is affirmative, we have the Hartree–Fock class of models for the F-center. A Hartree–Fock equation for the F-electron will be constructed, and its solution attempted using one of several approximations. These approximations themselves may be divided into two classes: the tight binding and the effective mass approximation. In the first class we find the point ion [26], the extended ion [27] and the pseudo-potential [28] approximations. In the second class we find the continuum and semi-continuum models. The neglecting of the electron correlation effects constitutes one of the fundamental limits of the Hartree–Fock approach; in fact, the polarization of the nearest electronic shells by the F-electron is due to such effect. The neglecting of correlations is not too important as long as the F-wavefunction is so compact that the electronic charge is completely neutralized by the positive effective charge of the vacancy. In fact, the calculations in the tight binding Hartree–Fock approximation cited earlier, yield results in reasonable agreement with experiment for the F-absorption. It is however possible to introduce partial polarization effects in the Hartree–Fock scheme by using effective potentials: in this type of model, the F-wavefunction is assumed to be extended, i.e. that polarization effects exist.

If the answer concerning the equivalence between the F-electron and the crystal electrons is negative, that is, if the F-electron is assumed from the beginning to be "different", one is brought to the so-called quasi-adiabatic approximation [29, 30]. Such an approximation is based fundamentally, on the observation that the F-electron binding energy (about 2 eV) is smaller than the Madelung energy (about 10 eV); the F-electron is thus slower than the electrons on the nearby ions. By analogy with the separation between nuclear and electronic coordinates ensuing from the adiabatic approximation, the quasi-adiabatic approximation means that the crystal electrons may follow adiabatically the motion of the F-electron and thus be polarized as the F-electron moves in a potential which will not depend on the positions of the nearby electrons (which however will appear as parameters). Thus, both in the Hartree–Fock and in the quasi-adiabatic schemes, the many-electron problem is reduced to a one-electron problem in an effective potential. However in the first approach it is difficult to treat polarization effects, whereas this is done automatically in the quasi-adiabatic approximation; on the other hand, in the latter method, the exchange correlations are totally neglected whereas in the Hartree–Fock approach they are naturally accounted for.

The form of the potential of the F-electron in the continuum model may provide an example of the different treatment of polarization effects in the two approaches. Since the crystal periodic potential is taken already into account by using the effective electron mass m^* instead of the real one, the potential seen by the electron will be that of the positive charge of the vacancy plus that due to polarization. In the continuum model, the polarization effects are described by the dielectric constant. If the nuclei are immobile, one must use ε_∞ (and ε_0 in the opposite case) [31]. Which is then the F-electron potential energy? We can see at once that we must use a self-consistent procedure, since the polarization potential is determined by the electron itself and by its position relative to the vacancy. The first contribution to the potential is due to the time-independent vacancy potential. Since the neighboring ions have an infinite time to adjust to this potential, we must use the static dielectric constant to describe the effect of the medium: the corresponding electric field will be:

$$-(1/\varepsilon_0)(\partial V/\partial r).$$

Now we must consider that a fraction of the electronic charge is always inside a sphere of radius r. This charge yields a field:

$$-(1/\varepsilon_0)(ep(r)/r^2),$$

where

$$p(r) = \int dr\, 4\pi r^2 |\psi(r)|^2 \qquad (4.17)$$

and $\psi(r)$ is the electronic wavefunction to be determined. In the quasi-adiabatic approximation, the electrons on the near ions may instantaneously follow the F-electron motion. In order to obtain the potential for this electron at point r, we may use the electron itself as test charge by placing it at point r; the ionic electrons will follow this motion, and thus we must add the term

$$ep(r)/\varepsilon_\infty r^2, \tag{4.18}$$

where the optical dielectric constant must now be used since the process is instantaneous and the ions cannot follow it. The F-electron potential energy at point r, is then given by

$$E(r) = -V(r)\frac{e^2}{\varepsilon_0} - Ce^2 \int_r^\infty \frac{p(r)}{r^2} dr, \tag{4.19}$$

where $C = (1/\varepsilon_\infty) - (1/\varepsilon_0)$ is the polaron coupling constant [32].

In the Hartree–Fock treatment of the same problem everything proceeds in the same manner with the exception that, since the electrons on neighboring ions cannot instantaneously follow the F-electron motion, expression (4.18) must be replaced by $ep(r)/r^2$, in which the optical dielectric constant is given the value of unity. Thus the Hartree–Fock potential energy of the F-electron is given by:

$$E(r) = -V(r)\frac{e^2}{\varepsilon_0} - \left(1 - \frac{1}{\varepsilon_0}\right)e^2 \int_r^\infty \frac{p(r)}{r} dr. \tag{4.20}$$

Both the Hartree–Fock and the quasi-adiabatic approaches lead to potentials which must be used self-consistently in a variational procedure. In general, it is natural to use the quasi-adiabatic approximation in continuum models, since this approximation becomes more valid as the electron becomes slower, i.e. as its wavefunction becomes more delocalized; this is of course the first condition for the validity of continuum models. A calculation along these lines was presented by Simpson [33]. His ground-state function for the F-electron is:

$$a_1(r) = (\mu^3/\pi)^{1/2}\exp(\mu r), \tag{4.21}$$

substituting this function in the quasi-adiabatic potential, he obtains:

$$E(r) = -(e^2/\varepsilon_\infty r) + (e^2/r)C(1 + \mu r)\exp(-2\mu r). \tag{4.22}$$

Then the variational wavefunction is introduced in the form:

$$a_2(r) = (\lambda^3/\pi)^{1/2}\exp(-\lambda r).$$

After minimizing the energy relative to the parameter λ, he obtains $\lambda = \mu$ from the self-consistency condition. The results for NaCl are: $(1/\mu) \sim 2\,\text{Å}$

and $E_0 = -1.52\,\text{eV}$, respectively for the extension of $a(r)$ and the ground-state energy. The calculated energy is of the right order of magnitude (experimentally it is 2.7 eV), however the discrepancy is too large. The calculation also leads to a very compact wavefunction, which is in disagreement with the requirement of an extended wavefunction for the continuum model. This contradiction is found in all effective mass models since they postulate the existence of a physically meaningful dielectric constant and it is by no means evident that this is the case in the vacancy, where the material medium does not exist. This inconsistency of the continuum model may be overcome in the "semi-continuum" approximation. In this case, one uses the effective mass approach where its validity can be reasonably assumed, i.e. outside the vacancy: inside the vacancy the potential is assumed to be that of a box. The semi-continuum model is thus a way of unifying the effective mass and tight binding approximations in a reasonable compromise. As all compromises, it generates new problems:

which is the proper dielectric constant to be used: ε_0, ε_∞, or an intermediate one?

what is the role of polarization effects in determining the potential depth inside the vacancy?

The answer to both questions may be found both in the Hartree–Fock and in the quasi-adiabatic approximations. The relative merits of these two ways of treating polarization have been discussed at length in the literature (see for instance the two particularly clear articles by Bennett [25]). Since the polarization effects are more important in emission, we shall limit our discussion to the relaxed excited states. In particular we shall discuss the work of Fowler on the relaxed excited state of the F-center in NaCl. Similar calculations have been performed by Wood and Joy [27], Kojima [27], Bennett [25], Iadonisi and Preziosi [35].

The semi-continuum Hamiltonian may be written as [34]:

$$H = (p^2/2m) + V_0, \qquad r < R,$$
$$H = (p^2/2m^*) - (e^2/\varepsilon_{\text{eff}})\,(1/r), \qquad r > R, \qquad (4.23)$$

where R is the vacancy radius and ε_{eff} an effective dielectric constant determined from the form of the ionic polarization potential. The constant V_0, representing the depth of the rectangular part of the potential, is given by:

$$V_0 = -(\alpha M/a) + W - \chi, \qquad (4.24)$$

where $\alpha M/a$ is the Madelung energy, χ the electron affinity of the neighboring ions and W a quantity which depends on the polarization. As we shall see later, both V_0 and ε_{eff} contain ψ, and thus it is necessary to use an iteration procedure to evaluate V_0 and ε_{eff}. Fowler assumed the electronic wavefunction in the 2p-like relaxed excited state to be very diffuse, and that therefore the electron was much slower than in the more compact groundstate; thus he chose the quasi-adiabatic approach to determine the polarization potential. Such a choice would lead to $\varepsilon_{eff} = \varepsilon_0$ for the extreme case of a very diffuse electronic state. In the more general case, ε_{eff} is calculated using the effective electron–hole interaction (with the hole effective mass set to infinity since in this case the "hole" is the vacancy) due to polarization, in the framework of the Haken theory of Wannier excitons. One then obtains:

$$-\frac{e^2}{\varepsilon_{eff}\, r} = -\frac{e^2}{\varepsilon_\infty\, r} - \frac{e^2}{r}\left(\frac{1}{\varepsilon_0} - \frac{1}{\varepsilon_\infty}\right)\left(1 - \frac{e^{-vr} + e^{-2r/a}}{2}\right)$$

where

$$v = [(2m/\hbar^2)\,\hbar\omega_{LO}]^{1/2} \tag{4.25}$$

a is the nearest neighbor distance and ω_{LO} is the frequency of the zone-center LO phonon. Then trial functions are chosen whose form is appropriate to the nature of the ground and excited states:

$$\psi_{1s} = \frac{\alpha^{3/2}}{(7\pi)^{7/2}}(1' + \alpha r)e^{-\alpha r}; \qquad \psi_{2p} = \frac{\beta^{5/2}}{\pi^{1/2}}r\,e^{-\beta r}\cos\theta; \tag{4.26}$$

where α and β, are variational parameters. The variational method then yields the energies of the 1s and 2p states and the values of the α and β parameters which determine the wavefunctions. Fowler has used the following values to obtain the transition energy for the 1s–2p absorption transition $a = 5.31a_0$ ($a_0 =$ Bohr radius), $R = 5a_0$ (Mott–Littleton radius), $\varepsilon_{eff} = 2.31$ and $\chi = 0.02$ a.u.; these numbers apply to the case of NaCl. In absorption, $\varepsilon_{eff} = \varepsilon_0$ since both ground and excited states are compact. The results are as follows:

$$\alpha = 0.56a_0^{-1}, \qquad \beta = 0.42a_0^{-1}, \qquad E_F = 2.80\ \text{eV},$$

where E_F is the absorption energy (which experimentally is 2.78 eV).

For the emission problem, Fowler has supposed that the nearest ions relax about 10% after the optical transition, so that $R = 5.5\,a_0$ and the contribution of the polarization to the V_0 part of the potential is not negligible the result is that V_0 becomes much less negative in the RES. Finally an appropriate value for $\varepsilon_{eff} = 4.2$. This value can be justified by a simple qualitative argument: when the electron is separated from the vacancy by a distance r larger than $1/v$, it moves slowly and thus the static dielectric constant is appropriate for $r < 1/v$ instead, the electron moves sufficiently rapidly that ε_∞ may be used. Now putting $m^* = 0.6m$ (this was Fowler's hypothesis which was experimentally verified later), one obtains $1/v = 26a_0$. Furthermore Fowler has obtained $\beta = 0.1a_0$ for the 2p-like function of F; this means that about 60% of the electronic charge is outside of the sphere of radius $1/v$ and 40% inside this therefore ε_{eff} must be given by

$$\varepsilon_{\text{eff}} = 0.6\varepsilon_0 + 0.4\varepsilon_\infty = 4.3,$$

in good agreement with the value obtained with a more rigorous calculation starting from (4.23). The emission energy was found to be 1.24 eV, and 0.12 eV for the ionization energy; both numbers are in good agreement with experiment. This is not surprizing since certain quantities (such as the ionic relaxation and m^*) have been chosen to fit the experimental data. However, as Fowler points out, these values are very reasonable and would have been chosen in this fashion anyway.

For F*, $\beta = 0.1a_0$ instead of the $0.42a_0$ of the unrelaxed excited state thus the wavefunction in the RES is much more extended than that of the unrelaxed state. The overlap between ground and relaxed excited state (RES) is thus smaller than for the case of the unrelaxed state. Fowler has calculated that because of this the transition probabilities ratio between emission and absorption is 0.13; this would explain the anomalously long decay time of the F-center emission.

The calculations by Fowler (and the similar ones that followed [35, 36], show that a simple and reasonable explanation of the emission properties of the F-center can be obtained by considering the polarization effects, and particularly the deformation of the electronic charge distribution which follows the ionic relaxation after absorption. Several authors have tried to give a more quantitative justification to Fowler's approach; for instance Wood and Opik propose a new method to evaluate the Hartree–Fock integral [24]. Their results quantitatively confirm that polarization effects must be considered in order to understand the properties of the F-center, both in absorption and *a fortiori* in emission. They also have shown that the semi-continuum model yields a good approximation for the determination of electronic states, particularly F*. Finally, in calculating the energy of the electronic states as a function of the ionic relaxation around the vacancy, they find that the 2s state, above the 2p state in absorption, ends up below the 2p state in the RES, with an energy separation of 0.08 eV. This result is important since it provides a theoretical basis to the model for F proposed by Bogan and Fitchen [37].

The 2s–2p mixing in the F relaxed excited state. The model of Bogan and Fitchen, based on Stark effect measurements in emission, starts from the following hypothesis: the RES of the F-center consists of a series of near-spaced levels whose wavefunctions have different orbital moments, and the transition probability to the groundstate of which is weaker in the lower energy states. In their simple model, Bogan and Fitchen have assumed that the 2s state is the lowest one, followed immediately by the 2p state. Furthermore these states are considered to be quasi-degenerate and strongly mixed by long wavelength LO phonons. Such vibrations are in fact associated with a macroscopic electric field, which mixes the 2s and 2p state as a sort of internal Stark effect, since the F-electron may follow adiabatically the ionic movements which cause the electric field itself. Assuming furthermore that the LO phonons involved are the T_{1u} symmetry modes, first-order perturbation theory yields, for a vibrational distortion in the x-direction

$$|2s'\rangle = (1 + \alpha^2)^{-1/2} (|2s\rangle + \alpha|2p_x\rangle)$$
$$|2p'_x\rangle = (1 + \alpha^2)^{-1/2} (|2p_x\rangle - \alpha|2s\rangle). \qquad (4.27)$$

The $2p_y$ and $2p_z$ states are not affected and remain degenerate. The internal Stark effect yields then the F* configuration shown in fig. 4.6. At low temperatures ($kT \ll \delta E$) only the 2s' state will be populated and emit, and the radiative lifetime of this state will depend on the mixing coefficient since

$$\frac{1}{\tau_R} \sim \sum_i |\langle 1s|x_j|2s'\rangle|^2 = \frac{\alpha^2}{1 + \alpha^2} M^2, \qquad (4.28)$$

where

$$M^2 = |\langle 1s|x|2p_x\rangle|^2.$$

The preceding equation shows clearly that the long lifetime of F* may be due to the predominantly 2s-like character of the emitting level. This

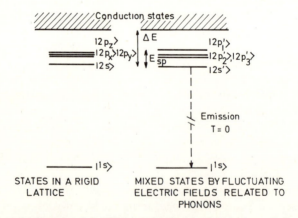

Fig. 4.6. Schematic diagram of the F*-energy levels: in a rigid lattice and after allowing for internal Stark level mixing (after Bogan and Fitchen [37]).

interpretation does not exclude, of course, that the emitting state also be diffuse according to Fowler's model. Actually, the very high values for α which have been determined experimentally [37] indicate that the predominance of the 2s character in the emitting state is not sufficient to entirely interpret its long lifetime. Within this model a simple explanation is found for the temperature dependence of the radiative lifetime. Temperature may influence this lifetime in two different fashions: first, by an increase of the parameter α due to a corresponding increase in the population of the optical phonons associated to the internal Stark field; secondly, by an increase in the populations of the higher levels of F*. The radiative lifetime of these levels is shorter since they are predominantly 2p-like. In fact, F* is certainly in thermal equilibrium since the luminescence decay may be described by a single exponential over four orders of magnitude of the intensity [38]: a thermally induced change of level population will then cause a change in the radiative lifetime.

Bogan and Fitchen have neglected the effect of phonon population change: the zero-point mixing is already so great (see table 4.4) that such an effect gives a negligible contribution in most of the temperature domain in which the decay time may be measured with some precision. In any case this is a very simplified model and one must be careful not to expect too much of it. Within this framework it is simple to estimate the temperature dependence of the decay time:

$$\frac{\tau(T)}{\tau(0\,\mathrm{K})} = \frac{1 + 3\exp(-\delta E/kT)}{1 + 3R\exp(-\delta E/kT)},$$

(4.29)

where

$$R = (1 + \tfrac{2}{3}\alpha^2)/\alpha^2$$

and δE is the mean splitting energy between 2s' and the three 2p' states, which for simplicity are assumed completely degenerate. The temperature

Table 4.4
Fundamental parameters for F* determined from fitting experimental Stark effect data with vibronic theory (From Imamaka et al. [41]).

	Δ	D_1	α^2	δE(meV)	$\hbar\omega < 0$(meV)
KF	0.0	0.5	0.37	13.1	41.4
KCl	1.25	0.35	0.15	16.3	28.6
KBr	1.50	0.25	0.10	14.9	21.0
KI	2.3	0.2	0.05	14.8	17.9
RbCl	1.5	0.3	0.12	15.0	22.3
RbBr	2.0	0.2	0.06	12.9	16.1
RbI	2.5	0.2	0.05	11.2	13.3

variation of τ was measured precisely for the first time by Stiles et al. [22]; the agreement with the preceding formulae is good especially in the lower temperature region, where the simple model is expected to work better in view of the approximations involved.

Within the framework of the Bogan and Fitchen model, an external electric field should decrease the lifetime of F* by causing an additional mixing of the 2s and 2p states. To evaluate this mixing, first-order perturbation theory may be used since the additional mixing $\Delta\alpha$ is certainly small relative to α itself. Summing over all three x, y, z possible distortion directions of the unperturbed states, the total relative variation of the radiative decay time turns out to be:

$$\Delta\tau_R/\tau_R = -(\Delta\alpha)^2/[\alpha^2(1 + \alpha^2)]. \tag{4.30}$$

This model predicts then, a quadratic decrease of τ_R with the intensity of the applied electric field; this decrease has effectively been observed for the F-centers in KCl and KF [23, 29]; more recently the Stark effect on F* has been also studied in KI, KBr, RbBr and RbI [40]; (fig. 4.7 and table 4.4).

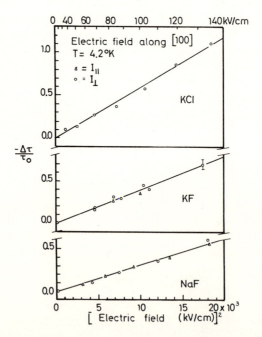

Fig. 4.7. The relative decay time change of the F-center in several alkali-halides as a function of the square of the applied electric field (after Stiles et al. [23]).

4.2.5. Recent spectroscopical evidences on F*

Many experiments have been performed to test the various models of F*. Historically, the diffuse state model was generally accepted at first, since a number of important experimental findings agreed with its predictions. Hetrick and Compton [42] studied the stress-induced linear polarization of the F-center emission. This induced polarization was found to be linear in the applied stress intensity. Furthermore, the effect was independent of temperature, emission wavelength, F-center concentration and incident light polarization. These last results show that the F* state is not measurably affected by the applied stress: if this were so, in fact, a temperature or wavelength dependent effect should have been detected. Hetrick and Compton have interpreted their results by considering the stress-induced mixing of the groundstate with higher lying d-states. In any case F* was taken to be extended since it was not appreciably perturbed by the applied stress.

Another experiment seemed to confirm the diffuse model for F*: the measurement of the decay time of the F_A-emission in KCl [43]. The logic of the experiment went as follows: if F* is extended, then even a large perturbation should not affect the decay time too much; in the opposite case, a substantial effect should be observed. Now the luminescence of the F_A(Na) center in KCl is very similar to that of the F-center, with a small shift of the emission towards lower energies: 1.12 eV for F_A(Na) as opposed to 1.24 eV for the F-center [44]. The F_A-emission decay time was found to be only 10% shorter than for the F-center. The conclusion was thus drawn that F* is extended. However, the important quantity to consider is not the radiative lifetime measured, but rather the transition matrix element, related to it by:

$$1/\tau_R \sim E^3 |M|^2, \tag{4.31}$$

where E is the emission energy. Thus a 10% variation in lifetime indicates in reality a 50% variation in M, given the emission energies involved. These lifetime measurements therefore do not exclude the possibility of a more compact predominant 2s-like state for F*.

The Bogan–Fitchen model was in any case strongly supported by Stark effect measurements. We have already mentioned the quadratic decrease of with applied field intensity [22, 45]. Bogan and Fitchen [37] and Künhert [46] have detected the Stark induced linear dichroism in the F-emission. Recently these measurements have been extended and generalized by Ohkura et al. [40] who also offered a theoretical interpretation based on vibronic theory [41]. The effect is due to the field-induced increase of transition probability for emission polarized parallel to the field and the corresponding decrease for the perpendicularly polarized emission. The dichroism signal also varies quadratically with the field intensity. The most important feature of these experiments is the variation of linear dichroism with temperature: it is constant below about 40 K and decreases exponentially at higher temperature. This behavior, shown for all the F-centers studied, can be simply interpreted only for a non-orbitally degenerate lowest level. Thus these experiments are in agreement with the predominantly 2s character of the F*-emitting state, as hypothesized by Bogan and Fitchen.

Very detailed information F* has been obtained with the study of magnetic effects and especially with the development of optical detection of magnetic resonances in F*, (to which we shall devote the next section). Fontana and Fitchen [47] first detected the diamagnetic part of the magnetic circular polarization of the F-center emission in KF, and later in KCl [48]. The signal and its temperature variation could be well interpreted on the basis of the Bogan–Fitchen model. More recently, both dia- and para-

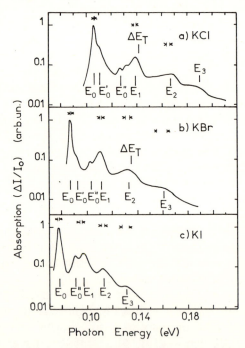

Fig. 4.8. Transient absorption spectra of additively colored crystals induced by F-band excitation at 10 K. The induced absorption has been normalized at the maximum value: ΔE_T represents thermal ionization energy of the F*-center. The previous results of Park and Faust on KI are shown in (c) for comparison. Spectral bandwidths in the experiments are indicated at the top of each spectrum (after Kondo and Kanzaki [50]).

magnetic components of MCP have been observed for KI and KBr [49].

Very detailed information on the electronic–vibrational structure of F* has been also obtained by Kondo and Kanzaki [50], who studied in detail (the first experiment of this type having been performed by Park and Faust [51]), the IR absorption spectrum due to optical transition from F* to higher lying states (fig. 4.8). It is important to note that F* is much more weakly coupled to phonons than F, as shown by the presence of a zero-phonon line and one-phonon sideband in the absorption spectrum (see chapter 3); furthermore, the level structure of F* is only compatible with a Bogan–Fitchen-type model.

Vibronic theories of F*. The experimental information discussed up to now spurred a notable theoretical effort to give a more quantitative basis to the heuristic models used to interpret the data. Basically the approaches have been of two types: the "polaron" approach, which in a sense follows the initial treatment by Fowler [34] and the "vibronic" approach which may be considered in the spirit of the original Bogan–Fitchen treatment. We shall limit our discussion to vibronic theories, which have been more detailed and thorough in their analysis and predictions. In any case, a recent bound polaron theory using a Hartree Hamiltonian has given results which are in reasonable agreement with some vibronic theories [52]. In particular it yields for the lowest level of F* a state which is 70% 2s-like and 30% 2p-like, with a radius of about 15 Å.

Vibronic theories are generalizations of the ordinary Jahn–Teller theory for degenerate electronic states to the case of quasi-degenerate states, or at least states the separation of which is not large compared to typical optical phonon energies.

Basically two treatments have been discussed, by Ham [53] and by Kayanuma and Toyozawa [54]. In both cases the F-electron is assumed to be embedded in a dielectric continuum of O_h symmetry and may interact with Γ_1^+, Γ_3^+, Γ_5^+ and Γ_4^- phonons (or local modes). Although both theories account for most of the experimental evidence, Ham's theory (and its subsequent application to prediction of experimental findings for F*) [55] does not predict the electric field induced red-shift of the F-emission band (Bogan and Fitchen [37]) and also leads to an unreasonably large splitting between the lowest and next lowest levels in F*. The theory discussed by Kayanuma and Toyozawa does not have these shortcomings and is also more general, since it treats the vibronic problem for all coupling strengths, whereas the Ham theory is limited to the weak coupling case. Thus here we shall discuss briefly only the Kayanuma–Toyozawa theory, inviting, however, the reader to peruse the elegant and in any case antecedent treatment by Ham and Grevsmuhl [55].

Using the local inversion symmetry of the F-center, and taking into account the loss of translational symmetry due to the presence of the defect, the Hamiltonian is written in terms of the spherical interaction coordinates for its vibrational part: these interaction modes however are still normal coordinates, with all their properties, for dispersionless vibrational modes. The Hamiltonian is further limited to the quasi-degenerate 2s–2p electronic subspace and its interaction with s-like, p-like and d-like modes; the effect of d-mode interaction is furthermore neglected since this interaction only causes the ordinary J–T effect, which does not mix even and odd states.

The "polaronic" effects of the electron–vibration interaction are not explicitly considered, although their effect is partially taken into account in the form of renormalization of wavefunctions and energy levels. The vibronic Hamiltonian thus obtained is then used to set up an infinite series of secular equations which are then truncated at the appropriate vibronic level for computer calculations which yield the energy level schemes as a function of some basic parameters: Δ, energy difference between the 2s and 2p levels before the electron–phonon interaction is turned on; S_0, the coupling strength to s-mode vibrations; S_1, coupling to p-mode phonons. The states corresponding to these levels are classified by the total angular momentum number J. In fig. 4.9, we show an example of such energy schemes. We have

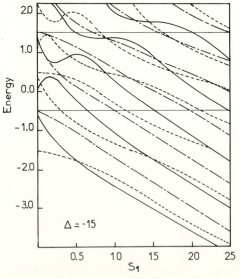

Fig. 4.9. Vibronic energy scheme versus p-mode coupling constant S_1 with Δ fixed to be -1.5 for $J^p = 0^+$ (both solid lines), I^- (broken lines) 2^+ (chain lines) and 1^+ (thin solid line). Only p-mode interaction is taken into account (after Kayanuma and Toyozawa [54]).

chosen on purpose that scheme in which the parameter values (in particular $\Delta < 0$) correspond to a situation where for zero coupling the 2p level is below the 2s level. Thus even in this case, for sufficiently strong coupling, the 2s-like level turns out to be the lowest level. Independently of the sign of Δ, for sufficiently strong coupling S_1, the function $E_j^n(S_1)$ is well approximated by:

$$E_j^n = \tfrac{1}{2}\Delta - 1 - S_1 + n + [J(J+1) + 1]/4S_1, \tag{4.32}$$

where n is the nth vibronic level with angular momentum J.

Using the vibronic energy level schemes and the wavefunctions thus calculated, quantitative comparison with experiment may be made [56]. As an example, in fig. 4.10, we show the comparison between experimental and theoretical results for the temperature dependence of the radiative luminescence lifetime in the KCl F-center. The better agreement is obtained using the general formula from the vibronic theory:

$$\frac{\tau(T)}{\tau(0)} = \frac{P_1(0,1) \sum_J \sum_n (2J+1) \exp(-\delta E_j^n / kT)}{\sum_J \sum_n (2J+1) P_1(J,n) \exp(-\delta E_j^n / kT)} \tag{4.33}$$

where $p_1(J, n)$ is the 2p electronic state fraction in the nth vibronic level with angular momentum J, and $E_j^n = E_j^n - E_0^1$ is the energy difference between vibronic levels. The agreement with experiment is also good at low temperatures for the simplified formula which can be obtained from (4.33) in the low temperature limit:

$$\frac{\tau(T)}{\tau(0)} = \frac{1 + 3 \exp(-\delta E/RT)}{1 + 3R \exp(-\delta E/RT)}, \tag{4.34}$$

where $\delta E_1^1 = \delta E$ and $R = P_1(1,1)/P_1(0,1)$.

This formula is very similar to that which is obtained in the simple Bogan–Fitchen model. This is reasonable since such a simplified model is in a way only a two-level vibronic model, and at low temperatures only the lowest vibronic levels will be populated.

Thus it seems that now the experimental phenomenology for the properties of F* is reasonably well interpreted theoretically. There are still some discrepancies however. For instance, the interpretation of ENDOR data on F* in KI by Mollenauer and Baldacchini [57] as indicating that the lowest level is orbitally degenerate (and thus p-like) still needs clarification. Furthermore, although the experiment of Kondo and Kanzaki previously discussed is clear evidence not only for the "vibronic" model of F*, but also for the intermediate coupling theory of Kanayuma and Toyozawa, recent measurements of IR absorption from F* [58] of the type performed by Kondo and Kanzaki have been used to claim evidence for the weak coupling theory of Ham and Grevsmuhl. Thus even at the reasonably quantitative

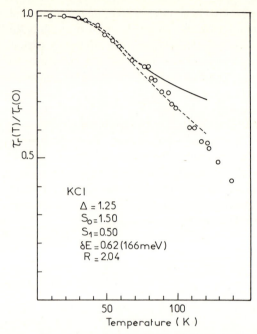

Fig. 4.10. Temperature dependence of the radiative lifetime of the excited F-center in KCl. The solid line and the broken line represent the best fit theoretical values. In the latter case, the lowest thirty vibronic levels are taken into consideration with $J^P = 0^+$, 1^- and 2^+. Experimental values are given in ref. [23] (after Kayanuma [56]).

level reached by theoretical and experimental understanding of the F-center, this old glorious workhorse of solid state physics does not cease to be surprizing and fascinating.

"Different" F-centers.   Still remaining within the family of the alkali-halides, there are some F-centers which do not behave according to stereotype. One for instance is the F-center in CsF (although in general all F-centers in cesium halides are different in the sense that the strong spin–orbit interaction of the cation yields bands which are somewhat structured); in this case the F-band shows a well defined structure, for the interpretation of which it is necessary to invoke the interplay of strong spin–orbit coupling and strong to intermediate J–T coupling (see for instance the refs. cited in Fulton and Fitchen [59]). Another atypical characteristic of the F-center in CsF is its anomalously short luminescence decay time (50 ns at low temperatures). Furthermore for this F-center there is a difference between the decay time and the groundstate repopulation time; these quantities should in principle be the same for an ordinary luminescence cycle. Thus some kind of bottleneck may exist during the relaxation of the F-center in CsF.

In recent years the F-center (or its analogs) have been detected and studied in host crystals not in the family of the alkali-halides. It would be out of place here to summarize all that has been done in this field. In any case, the F and F^+-center in the alkali-earth oxides have received particular attention. Since the host crystal anions are divalent, the F-center is a two-electron system and the F^+ contains one electron. Apart from these basic differences, in general the F-centers in these systems are less strongly coupled to vibrations, and thus their optical bands show vibrational features, including zero-phonon lines. The F- and F^+-centers in CaO have been particularly well studied by the Grenoble group [60] and by Harris et al. [61]. The simultaneous presence of a not so strong electron–vibration coupling and degenerate electronic states has made these systems ideal for the study of J–T interactions. Also for these systems, the entire sophisticated apparatus of excited state magnetic resonance spectroscopy has been used, with good and interesting results.

4.2.6. Optical detection of magnetic resonance in the F-center

The spin temperature of a spin system may be varied by phonon or microwave absorption, as we have seen in chapter 1; thus magnetic circular dichroism (MCD) or Faraday rotation may yield a sensitive technique for the detection of electron spin resonance. The first experiment in this direction was attempted by Karlov et al. [62], who displayed that the T_1 relaxation time and the effect of optical pumping of the Zeeman sublevel populations could be determined optically.

4.2.6.1. The determination of T_1. The relaxation time T_1 has been measured using the MCD of the F-center in the alkali-halides [62, 63, 64] and for some rare earth impurities in the alkali-earth fluorides [65]. At low temperature ($T \leq 5$ K) the spin relaxation is due to one-phonon processes since T_1 varies as $1/T$ in both types of systems studied [66, 67]. In order to measure T_1 the spin sublevel populations must be modified using microwave absorption, or a rapid variation of magnetic field, or optical pumping. We shall discuss this last method in more detail later on. If the external magnetic field is suddenly brought to zero, the spin polarization will relax back to its equilibrium value with the characteristic time T_1. Now this polarization is connected with, and in some case proportional to, the MCD signal; thus T_1 may be simply measured by following the time evolution of the MC signal S after a sudden variation of the magnetic field. By varying H from a high value to a given lower value one then measures relaxation time for the end value of the field; the $T_1(H)$ behavior may be thus obtained [63]. This method is of course only applicable for long relaxation times. For short times, the spin population may be brought out of its equilibrium distribution using optical pumping, and the resulting MCD variations may be followed on an oscilloscope [63]. When applicable, the MCD technique for determining T_1 has some important advantages:

it is extremely sensitive and thus T_1, which may vary strongly with concentration can be measured in dilute systems [63, 67];

it is not necessary to use microwaves, which simplifies the experimental hardware and also allows measurements at high fields (~ 120 kG), i.e. for widely split Zeeman sublevels.

The results of Panepucci and Mollenaur on T_1 for the F-center in KCl, KBr and KI are reported in fig. 4.11; they were able to show that the main relaxation coupling mechanism was due to the phonon-induced modulation of the hyperfine interaction between the F-center and its nearest neighbors [68]. They have also observed that in weak fields ($H \leq 7$ kG) that the relaxation is due essentially to inter-center interaction and that above 25 kG the phonon modulation of the spin–orbit interaction also contributes significantly to the relaxation [69].

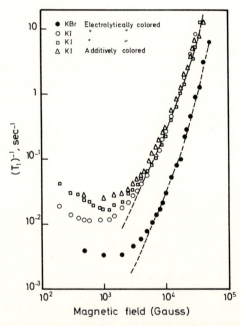

Fig. 4.11. Variation of the relaxation time T_1 of the F-center optically determined at 1.65 K as a function of magnetic field in KI and KBr (after Panepucci and Mollenauer [63]); the dashed lines correspond to the theory discussed in this paper.

4.2.6.2. The optical pumping.   For the F-center, the optical pumping cycle
goes as follows: there is first the absorption due to the 1s–2p transition;
thereafter the electron state quickly relaxes to the F*-state, yielding
about 1/4 of the absorbed phonon energy to the lattice via phonons.
After a characteristic lifetime F* emits a photon returning to the electronic groundstate, which then relaxes back to the initial state by phonon
emission. In a magnetic field the spin system will be polarized following
eq. (4.35):

$$\langle S_z \rangle = A \tanh (g\beta H / 2kT).\tag{4.35}$$

What is then the effect of the pumping cycle on the spin polarization?
McAvoy et al. [70] have answered that the polarization must be lost totally
due to the strong electron–phonon coupling relaxation, i.e. the spin
memory during an optical cycle must be zero (see also ref. [72]). More recent
results have shown that this is wrong. Schmid and Zimmerman [71] have
measured the EPR signal variations due to optical pumping intensity for the
F-center in KCl: their results show that the spin memory per cycle was larger
than 95%. Their results were later confirmed by Mollenauer et al. [72],
Fontana [64], Porret and Lüty [73]. These authors have independently
shown that spin memory per cycle is larger than 99% in KCl and KF.

   In Fontana's experiment, the F-center spin polarization was detected
by the F-band MCD with a field of 15 kG at 1.3 K. Under the action of
F-band pumping light, the spin may relax either following a normal
Van Vleck one-phonon process or after possible spin memory losses
during the optical cycle. Thus the total relaxation rate $1/\tau$ is given by
[62, 71, 72]:

$$(1/\tau) = (1/T_1) + \varepsilon u,\tag{4.36}$$

where ε is the spin memory loss per cycle and u, the number of cycles per
F-center per second. If $\varepsilon u \gg 1/T_1$ the relaxation will be due essentially to the
optical pumping: if $\varepsilon u \sim 1/T_1$ both contributions to $1/\tau$ can be determined by
plotting it versus the pumping light intensity, i.e. u. Fontana did observe a
linear behavior from whose slope and intercept, ε and $1/T_1$, could be respectively determined. For KF, $\varepsilon \sim 0.01$. It is very surprising to find such a small
spin memory loss during a cycle in which about half of the absorbed energy
is lost by coupling to the lattice. In any case the small loss is finite, and thus
a sufficiently intense light might quench the spin polarization as the reader
may observe in fig. 4.12, where the solid curve represents the prediction form
eq. (4.12), and no free parameters were used. Thus optical pumping may
have considerable effects on the spin polarization, even if the spin memory
per cycle is practically total.

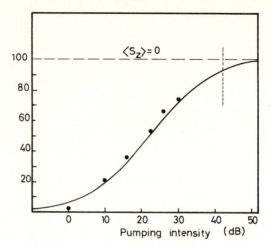

Fig. 4.12. Relative attenuation at 1.3 K of the paramagnetic component of MCD signal of the F-center in KF as a function of pumping light intensity. The full line is the theoretical curve with $T_1 = 90\,\text{s}$ and $\varepsilon = 0.01$ (after Fontana [64]).

4.2.6.3. EPR on F*. The MCD technique for EPR detection is very sensitive; so sensitive in fact that it made possible the detection of paramagnetic resonance in the relaxed excited state of the F-center [72], a remarkable achievement if we consider that the F*-lifetime is of the order of 10^{-6} s. In these experiments Mollenauer et al. have used the good spin memory efficiency of the F-center: furthermore they have used circularly polarized optical pumping. The pumping wavelength was chosen to coincide with one of the maxima of the MCD signal, for which $(u^+ - u^-)/(u^+ + u^-)$ is maximum, where u^+ and u^- are the pumping rates for circularly polarized light from the $-1/2$ and $+1/2$ Kramers levels respectively. Assuming now that the spin–lattice relaxation time in the relaxed excited state T_1 is much greater than the F* lifetime and that the pumping rate is small relative to the emission probability, Mollenauer et al. [74] have shown that the pumping rate from the groundstate is:

$$p_1 = \frac{n^+ - n^-}{n^+ + n^-} = -\frac{1}{1 + (T_p/T_1)}\left[\frac{u^+ - u^-}{u^+ + u^-} + \frac{T_p}{T_1}\tanh\left(\frac{g\beta H}{2kT}\right)\right], \quad (4.37)$$

where $n^{\pm}$ are the $+1/2$ and $-1/2$ level populations respectively and $1/T_p$ the relaxation rate due to optical pumping and microwave absorption in F* (when microwave power vanishes T_p^{-1} reduces to the term εu of eq. (4.36). If there is no optical pumping:

$$p_1 \to -\tanh\left(g\beta H/2kT\right), \quad (4.38)$$

and in the case of optical saturation

$$p_1 \rightarrow - (u^+ - u^-)/(u^+ + u^-). \tag{4.39}$$

In this last case the MCD signal is independent of the applied field. Even though this signal may be sizeable (5% in KCl and 40% in KI) by itself, circularly polarized optical pumping may increase it over its equilibrium value for a given field and temperature (in this way T_1 for very weak field may be measured by using the pumping enhanced signal [62, 63]). If pumping is reduced from the saturation level and if $T_p \sim t_1$, p_1 will be very sensitive to variations in T_p which in turn may be caused by microwave absorption in the excited state: thus the excited state resonance may be detected by measuring the groundstate spin polarization. In fig. (4.13), we report typical results obtained by Mollenauer et al. and in table 4.5 the relative g and g^* values obtained.

4.2.6.4. Optical detection of ENDOR in the ground and relaxed excited state of the F-center. The degree of sophistication of the preceding experiments has been increased and ENDOR in F* has been detected. In order

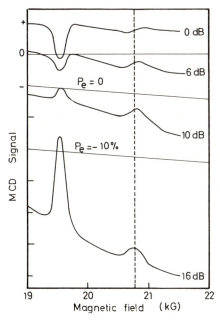

Fig. 4.13. Optical detection of the ESR of the F-center in KBr at 1.3 K for different intensities of the pumping light. On the left, resonance of the groundstate; on the right, on the dashed line, resonance of the relaxed excited state F* (after Mollenauer et al. [72]).

Table 4.5
g-values for the ground and relaxed excited states of the F-center in several alkali-halides.

	g(ref. [66])	g(ref. [77])	g^*(refs. [72, 74])	g^*(ref. [77])
NaCl	1.997	1.999		1.970
KF	1.996	1.994		1.995
KCl	1.995	1.984	1.976	1.980
KBr	1.982	1.982	1.862	1.873
KI	1.964	1.968	1.627	1.631
RbBr	1.967	1.988		1.844
RbI	1.949	1.953		1.630

to detect ENDOR, the sensitivity of the measurement had to be increased and for this modulation techniques had to be employed to their utmost. Since T_1 is long, it had to be artificially decreased in order to use lock-in amplifiers; this was done using saturated optical pumping. From eq. (4.36), T_1 is inversely proportional to the pumping intensity, which may be sufficiently intense to bring T_1 down to some milliseconds. The spin system has then very little inertia and may follow modulation up to a frequency of 1 kHz. Modulation was twofold; first the polarization was varied between σ^+ and σ^- at a frequency of 50 kHz. For saturated polarized pumping, the groundstate spin polarization is given by eq. (4.37). For AC pumping at high frequency, $p_1 = 0$, and the spin polarization is completely quenched. If now a radiofrequency or microwave field which may induce transitions is applied when, for instance, the pumping light is σ^+ the pumping rate will be higher for σ^+ than for σ^-. The equilibrium between the populations will be altered and $p_1 \neq 0$. The second modulation is achieved by applying the radiofrequency and microwave fields when the pumping is σ^+ or σ^- respectively and at a frequency sufficiently slow (some tens of Hz [74]) that the spins will be able to follow and p_1 may then be detected with great sensitivity.

Using microwaves, the EPR signals for the ground and the relaxed excited states have been obtained with much greater precision than with the MCD measurements discussed before. If now the EPR line is satured and a modulated radiofrequency field is applied the ENDOR spectrum of the ground or excited state may be obtained, depending on which line is saturated. Here we must recall that the groundstate ENDOR spectrum had already been obtained optically by Parry et al. [75] with a somewhat similar method; we refer the reader to this work for details. The ENDOR spectrum obtained by Mollenauer et al. is shown in fig. 4.14 for the case of F* in KI. Using the technique described below, Schwegg et al. [78] were also able to detect ENDOR on F*.

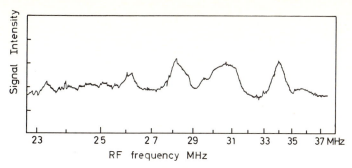

Fig. 4.14. Optical detection of the ENDOR spectrum of the F* state in KI (after Mollenauer et al. [74]).

4.2.6.5. Concentration quenching of F-center luminescence and optical detection of magnetic resonance.

Another and particularly elegant method for optical detection of EPR, and also ENDOR, in the F-center has been introduced by the Neuchatel group [76, 77, 78]. It is based on the F-spin polarization variation connected with the luminescence concentration quenching. It is well known [79] that the F-center luminescene yield η varies with the concentration c; at a sufficiently elevated concentration, η decreases exponentially with c, which indicates that the quenching is due to inter-center interactions. Porret and Lüty [73] have interpreted the quenching mechanism: the electron of an excited F-center may tunnel to another F-center to form F^-. The electron may then return to the F-center it left, yielding F in its groundstate: the net results is the radiationless de-excitation of that F-center. The formation of the intermediate F^--centers plays then an important role in the luminescence yield (for $c \sim 2 \times 10^{16}$ F cm^{-3}). Since the only stable state of F^- is a singlet [81, 82], η will be very sensitive to the spin polarization; the F^--center would in fact only form if the two electrons have antiparallel spins [82].

The effect is easy to understand: if the spin polarization is total, the F^--centers may not be formed since the electron spin memory is almost total, and the electron will retain its original spin. Since it cannot leave F*, it will return to the groundstate via radiative emission. Thus spin polarization and spin memory may completely annul the luminescence quenching. Porret and Lüty have effectively verified this, by measuring the MCD and using excitation intensity for the luminescence so that polarization was not disturbed from its value [73]. They measured the luminescence intensity at 1.7 K as a function of magnetic field up to 90 kG and obtained the results of fig. 4.15. For a sufficiently high field, the quenching of the luminescence caused by tunnelling is practically neutralized. In the figure, the dashed curve repre-

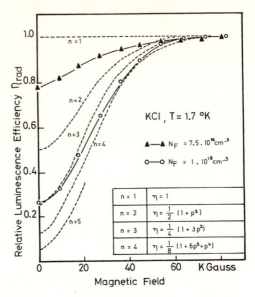

Fig. 4.15. Variation of the F-center luminescence yield at 1.7 K as a function of the applied magnetic field in two KCl crystals with different F-center concentrations. The dashed curves represent the behavior predicted for agglomerates of $n = 1, 2, 3, 4$ F-centers (after Porret and Lüty [73]).

sents the results of a simple statistical model for the F-center distribution in the crystal: the numbers 1, 2, etc. correspond to the number of interacting F-centers. These results then yield precious information on the F-center distribution in the crystal. This effect may also be used to detect the EPR of F-centers (or rather F-center "pairs") since the luminescence yield is directly connected with the electron spin polarization. If a microwave field is now applied to the crystal so that transitions between the Zeeman sublevels are induced, the spin polarization will diminish, and this will cause a diminution of η.

This signal is easier to detect (it involves the measurement of an intensity) than the MCD signal. The diminution of luminescence intensity at resonance was observed by Ruedin and Porret in their first experiments [76]: the results of Porret and Lüty have yielded an interpretation for this decrease. This technique for the detection of EPR may be used to study the resonance in the ground and in the excited states: it has been applied for a detailed study of F* [77, 78]. In particular it was shown that above 20 K, the spin relaxation in F* obeys an Orbach mechanism. This result is in good agreement with the Bogan–Fitchen model for F* since the Orbach energy is found to be very close to the 2s–2p splitting as determined by Stiles et al. [22, 23]. Ruedin et al. [77] have also precisely determined g and g^*. These results are reported together with those of Mollenauer et al. [66] in table 4.4. The precise knowledge of g and g^* and the use of the Bogan–Fitchen wavefunctions allow the evaluation of the spin–orbit coupling in F*. First-order perturbation theory yields:

$$g^* = g\{1 - \tfrac{2}{3}[\alpha^2/(1 + \alpha^2)]\,(\lambda/\delta E)\}, \tag{4.40}$$

where λ is the spin–orbit coupling constant, α the mixing s–p coefficient and δE the relative splitting. Since all the quantities in this equation are known, λ may be determined, with an error mainly influenced by the error on α. For KF, $0.03 < \lambda < 1.4$ meV and for KCl $0.4 < \lambda < 1.4$ meV. Even if qualitative, this result is nevertheless very important since it shows that there is a quenching of the spin—orbit interaction in F* (in KCl, the λ value for the unrelaxed state determined by MCD absorption measurements is 10 meV). This is a manifestation of the so-called Ham effect we have discussed in the preceding chapter.

The many sophisticated experiments performed on F and F* have yielded enough precise and stimulating experimental information that in recent years a new theoretical effort has been devoted to the understanding of F*-electronic states. Here we shall conclude by mentioning that, in principle, the type of detailed information obtained with the simultaneous use of microwave and optical techniques for the F-center should be also obtainable for other defect electronic systems; as we shall see later in this chapter, good results along these lines have been obtained for alkali-halides phosphors.

4.3. The exciton in the alkali-halides: propagation and self-trapping

It could appear quite inadequate to treat now, in this chapter, the question of excitons in alkali-halides. In fact, as will be shown in this section, excitons are quickly self-trapping after they have been created by light excitation and then they form a localized electronic state which can be described like a molecular impurity substitutional in the lattice.

At low temperatures, alkali-halides excited in different ways (X- and gamma rays, electrons, UV) shows a characteristic luminescence [83]. The emission spectrum of NaI excited by X-rays, is shown in fig. 4.16; two main features are present, namely a sharp and weak peak very close to the absorption edge and an intense and broad band at about 4.2eV[84, 85]. Lushchick and co-workers have recently observed the sharp peak in a large number of ionic materials, whereas the broad emission has been known for a long time. Such broad emission may consist of one (NaI), two (KI) or several (RbI) emission bands. In table 4.5, are reported the main characteristics of the intrinsic luminescence in all the alkali-halides investigated. As we shall see,

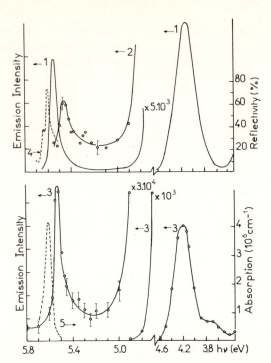

Fig. 4.16. Excitonic recombination fluorescence of NaI. (1), Electron excitation at 67 K; (2), UV excitation at 80 K; (3), X-ray excitation at 4.5 K; (4), reflection spectrum at 66 K; (5), absorption spectrum at 10 K; (after Lushchick et al. [84]).

Table 4.6
Energy and decay time of emission bands associated with the recombination of the self-trapped excitons in alkali-halides at low temperature.

System	Transition (A)		Transition (B)		Ref.
	E(eV)	10^{-9}s	E(eV)	10^{-6}s	
LiF	5.80	—	3.53	—	(a)
NaCl	5.6	≤ 5	3.38	295	(a)
	5.47		3.47	310	(b)
KCl	—	—	2.32	5000	(a)
	—	—	2.54	5000	(b)
RbCl	—	—	2.27	5500	(a)
	—	—	2.41	8900	(b)
NaBr	—	—	4.60	0.49	(a)
	—	—	4.65	0.46	(b)
KBr	4.42	9	2.27	130	(a)
	4.40	—	2.44	100	(b)
RbBr	4.20	11	2.10	180	(a)
	4.13	—	2.36	150	(b)
NaI	—	—	4.20	0.1	(a)
	—	—	4.24	0.57	(c)
KI	4.15	≈ 9	3.34	4.4	(a)
	4.13	—	3.31	6	(b)
CsF	5.96	—	3.13	—	(a)
CsCl	5.07	—	4.52	—	(a)

(a) Ref. [87]. (b) Ref. [86]. (c) M. P. Fontana, H. Blume and W. J. Van Sciever, Phys. Stat. Solidi. **29** (1968) 159.

the variety of behavior in the various alkali-halides can be interpreted by a relatively simple model.

The main interest of the intrinsic luminescence is that it is indeed intrinsic, i.e. it is directly related to the return of the excited crystal to its groundstate. The excited electronic states may be either excitonic states or free electron and hole states. Thus the intrinsic luminescence will yield information on these states. Since the free electron and hole pairs, which are not trapped otherwise, might trap each other through the long range Coulomb interaction to form an exciton, the ensuing luminescence will not be different from the direct excitonic emission; this is in fact what is observed and therefore it will suffice to study only the excitonic luminescence.

4.3.1. The self-trapped exciton

In the alkali-halides the hole effective mass is practically infinite, i.e. in a time of the order of a lattice vibrational period an inter-band transition hole will be trapped by a potential caused by its own presence: the hole will self-trap. The corresponding "defect" called V_k-center has been discussed in some detail in chapter 1 (§ 1.1.2.1.). It is reasonable that the very large effective mass of the hole will slow down the exciton until it will stop to become the self-trapped exciton. In other words, the self-trapped exciton will consist of a V_k-center around which the electron revolves in a more or less tight orbit. The intrinsic luminescence will then be due to the electron–hole recombination which will take place only after the excitonic self-trapping. The self-trapping time is not vanishing however and thus it is possible that the excitonic recombination takes place while the exciton is still free: in this case the emission would take place with little Stokes shift, i.e. very close to the absorption edge, and would be reasonably sharp, by analogy with the weakly coupled excitonic emissions in semiconductors; in NaI, the weak and sharp peak shown in fig. 4.14 is due to the free exciton luminescence, which then can be used to study exciton motion in the crystal; we shall discuss this aspect of the problem later on.

Returning now to the better known broad intrinsic luminescence we shall discuss an experiment which most conclusively connects it to the self-trapped exciton [86]: a crystal is irradiated at low temperatures with X-rays to create F- and V_k- centers. The V_k-center is anisotropic (D_{2h} $\langle 110 \rangle$ symmetry), and it is possible to align it optically (see § 1.1.2.1.). In a crystal in which the V_k-centers are aligned in a given $\langle 110 \rangle$ direction, the F-center may be ionized to yield free electrons which will eventually recombine with the V_k-centers; the ensuing emission being spectrally identical to the excitonic emission, is polarized along the orientation direction of the V_k-centers.

The self-trapped exciton features several optical properties which we shall presently discuss. In most cases the intrinsic emission consists of two emission bands A and B. These bands have different properties: the high energy emission (A) is polarized in parallel to the self-trapped exciton axis (σ emission) whereas the B band is perpendicularly polarized (table 4.5). Their decay times are also different. By excitation with 20 ns electron pulses Kabler and Patterson [86] were able to measure the luminescence decay times, τ. For the B band, τ is long and decreases rapidly with increasing temperature [87] with a law of the type:

$$\tau(T) = c \exp(-W/kT), \tag{4.41}$$

where W is 7.3 meV for KCl, 85 meV for NaCl and 110 meV for KI, for instance. The V_k–e center may be treated as a X_2^{-2} rare gas molecule in D_{2h} symmetry; such a molecule has spin singlet and triplet states and thus the B-band transition would be due to a forbidden triplet–singlet transition; the A band would instead be due to a singlet–singlet transition.

In order to obtain a more quantitative description of the triplet state, Fuller et al. [88] have detected the excited state absorption ensuing after pumping into the triplet state with electron pulses (fig. 4.17). By measuring the time decay of the pumping induced optical absorption bands, they have been able to connect them without ambiguity to the triplet state of the self-

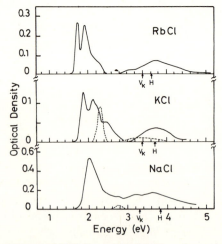

Fig. 4.17. Absorption spectra of STE formed by electron pulses at low temperature. Wavelength of V_k- and H-center absorption bands are indicated by arrows (after Fuller et al. [88]).

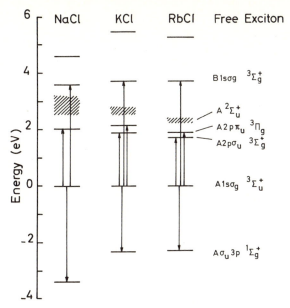

Fig. 4.18. Energy levels corresponding to main transitions from the STE ground triplet state $^3\Sigma_g^+$ in several alkali-halides. The bottom of the conduction band measured optically is represented by the dashed part (after Fuller et al. [88]).

trapped exciton. The excited state absorption consists of two groups of bands, for the case of Cl_2^{-2}: one band at 350 nm (about 3.8 eV) and one or more bands in the region of 700 nm (2 eV). The absorption characteristics of the triplet state may be interpreted with the energy level diagram shown in fig. 4.18 where the shaded area indicates the conduction band minimum before the lattice relaxation connected with the exciton self-trapping.

The triplet–singlet transition at low temperatures is quite complex and involves two different mechanisms. As stated before, the B-band decay time decreases exponentially with increasing temperature; this is probably due to a non-radiative transition between the triplet state and a singlet state lying slightly above it (approximately 15 meV for KCl [89]). However a more detailed analysis of the STE decay shows two components at low temperatures [90, 91]. For example, in KCl the 2.28 eV emission excited by two-phonon absorption of a pulsed N_2 laser light shows at 1.8 K a fast (100 μs) and a slow (3 ms) decay. The latter one varies rapidly and exponentially with increasing temperature. This behavior can be understood as follows: the $^3\Sigma_u^+$ state is split in the crystal D_{2h} symmetry into three levels B_{2u}, B_{3u} and A_u as shown in fig. 4.19. The transition from the lowest A_u level to the singlet state is strictly forbidden whereas the transition from the B_{2u} and B_{3u} levels is partly allowed by spin–orbit mixing with the higher singlet

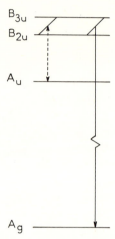

Fig. 4.19. Schematic energy diagram of the STE triplet groundstate and neighboring singlet state. (from Fischback et al. [90]).

state: thus the only way for the A_u state to relax is to be thermally excited to the B_{2u} and B_{3u} states. The zero-field splitting E_{ba} between the A_u and the B levels can be determined by this population effect and is found to be very small: 0.694 meV in KI [90] and 1.5 meV in KBr [91]. The spin triplet character of the STE has been confirmed by the optical detection of its electron spin resonance spectrum. Marrone and Kabler [92] have shown that the lowest emission band exhibits a partial circular polarization in a magnetic field, which has been subsequently used for ESR detection of the STE during continuous X-ray excitation [93, 94]. From these experiments, it was concluded that E_{ba} is effectively small, in good agreement with the result obtained from the temperature behavior of the luminescence decay times.

To summarize these different results, it can be concluded that the STE may be described as a hole strongly localized on two neighboring halogen ions in a $\langle 110 \rangle$ direction forming an X_2^- with an electron losely bound to it [95]. Its different energy levels could be described as excited states of the electron or of the hole. We shall discuss in chapter 5, the mechanism of color center formation by excitonic recombination: as we shall see, it is not yet clear if the photochemical reaction comes from an excited hole or from an excited electron state.

4.3.2. Excitonic self-trapping and energy transfer

The exciton is a mobile quantum of energy and this is its most important characteristic. It must then be possible to transfer energy across a crystal via excitons. The possibility of such transport has considerable implications for the understanding of energy fluxes in biological systems. Such systems are much more complicated than an alkali-halide crystal, and thus a quantitative explanation of the photo-synthetic process for instance has not been given yet. It is thus interesting to study energy transfer in simple systems. Most studies in the subject have been performed using the phenomenon of host sensitized luminescence [83, 96]: the effect consists in the observation of impurity luminescence upon excitation in the *intrinsic* absorption region. Hattori et al. [97] studied exciton diffusion in $RbI:Tl^+$. The luminescence of an exciton recombining on a substitutional Tl^+ site is easy to observe in this system and is different from ordinary exciton recombination luminescence. The authors observed that the intrinsic luminescence decreases and the Tl-related luminescence increases, at a given temperature, with Tl^+ concentration while the total yield for both remains constant. It is concluded that the Tl^+-induced luminescence is due to exciton diffusion and that the ratio of the two emission intensities can yield the diffusion coefficient, which is shown to vary according to the following equations:

$$D \sim T^{-1/2}; \qquad T < 20 \text{ K}$$

$$D \sim T^{-1/2} [\exp(\Delta E/kT) - 1]; T > 20 \text{ K}. \tag{4.42}$$

Instead of analyzing the final products of excitonic relaxation (on a perfect part of the crystal or at an impurity), it is also possible to work on the unrelaxed exciton which is observable through the sharp higher energy peak shown in fig. 4.16. Hayashi et al. [98] have looked in detail at the thermal behavior of the unrelaxed excitonic emission in KI and RbI. They observed that the intensity of the peak varies as follows:

$$I = I_0 [1 + c \exp(-\Delta E/kT)]^{-1} \tag{4.43}$$

similar to eq. (4.42); the same activation energy was also measured. From this they concluded that the exciton may relax either radiatively or non-radiatively, with the non-radiative part being governed by an activation energy as proposed earlier by Sumi and Toyozawa [99]. Similar results have been obtained by Guillot et al. ref. [100] in RbI excited with UV light instead of electron excitation; they have shown that the energy difference between the first excitonic peak and the free excitonic emission is only 2 meV. It is not clear to us whether this activation energy corresponds to a non-radiative decay or to a phonon assisted mobility of the free and unrelaxed exciton as the previous experiment would suggest. In any case, the exciton does seem to be mobile and can transfer energy before self-trapping; furthermore the self-trapping will take a finite time since in two-phonon excitations (in a KI crystal at 77 K) with a pico-second laser, Suzuki and Hirai [101] have observed that the triplet STE is formed within a time of about 200 ps. This last result indicates that the system needs time to relax to the STE triplet state.

4.4. Resonant electronic states

4.4.1. *Theoretical preliminaries*

The introduction of the defect in a perfect crystalline lattice leads to resonant or localized vibrational modes (chapter 2). The same may happen for electronic states, and in the preceding part of the chapter we have discussed in some detail the localized electronic states associated with the F-center; the approach we have used was essentially a tight binding approach, i.e. we have considered the F-center as a "molecule". However, since the F-center has no intrinsic degrees of freedom, from a conceptual point of view it would be more satisfactory to obtain the electronic states directly from the band structure of the perfect crystal. In practice this approach is not particularly useful in the localized state of the F-center; it becomes necessary however when treating the higher excited states, and in particular the states giving rise to the so-called L-bands. Later on we shall see which experimental evidence leads to the conclusion that such states are located deep into the conduction band of the crystal. This basic fact leads to some conceptually important questions:

(1) Why an absorption band is observed in connection with these levels?

(2) What is the fate of the electron thus excited?

The same kind of "discrete" states degenerate with the conduction band continuum have been observed in the excitonic spectra of semiconductors containing donor and acceptor impurities [102, 103]. In all these cases, the spectra are interpreted by assuming that the electronic states of the defect are formed from the band states of the pure crystal [104]. From a theoretical point of view, scattering theory constitutes the best method to approach this problem. It is well known for instance in nuclear or high energy physics that there are resonances in the scattering cross-section as the incident particle energy approaches some energy level of the target [105]. Such resonances are predicted by the simple Breit–Wigner formula.

Bassani et al. [106] have laid the rigorous theoretical foundations of the problem of defect-induced resonant and localized states. In particular they have shown that resonant states may appear that are related to minima in the band structure away from the Brillouin zone center. Optical transitions involving large wavevectors become possible since the wavevector conservation law does not hold as the translation symmetry is broken by the presence of the defect. Bassani et al. have essentially shown that scattering theory may be used in a periodic potential. The scattering potential due both to the defect and to the band structure of the crystal can be used to predict the existence of resonant modes. Such states will be built from Bloch functions in all reciprocal lattices or, if the reduced zone scheme is adopted, from all the bands. The resonant states connected with band minima at zone-edge point in the BZ may be strongly peaked (i.e. localized), and have lifetimes that are much longer than ordinary scattering states. The calculations of Bassani et al. lead to another interesting result: the transition probability from the groundstate to such resonant states should increase as the energy increases; this is because the delocalization of the resonant state in phase space increases with energy.

The existence of resonant states connected with secondary minima in the conduction band has been shown experimentally in doped semiconductors [103]; the L-bands in the alkali-halides seem to be due to the same mechanism. In fact, band structure calculations [107] have shown that there are supplementary minima in the band structure of alkali-halides for $k = 2\pi/a$, in the $\langle 100 \rangle$ direction.

4.4.2. Experimental evidences on the L-bands

The L-bands have been detected first by Lüty in KCl [108] in fig. 1.6 we show the absorption spectrum in their spectral region. There are three weak bands L_1, L_2, L_3 in order to increasing energy. The associated oscillatory strength is of the order of 0.01, i.e. the transitions are only weakly allowed. The oscillator strength seems to increase with light energy in agreement with the theory of Bassani et al. Excitation in these bands yields F-center ionization with almost unit efficiency, independent of temperature [109]. This is in agreement with the L-levels being degenerate with the continuum. It is however interesting to note that about 1% of the absorbed phonons are re-emitted as F-center luminescence [109]. Now, is the L-excited F-center luminescence due to the recombination of the ejected electrons with the F^+-center they left behind or does it come from the F-state, after crossing of the L-levels with the F-band level during relaxation? In other words, are all the F-centers ionized by L-band excitation, or is there a residual 1% that is not ionized? The answer to this question is important since it may yield information on the lifetime of the L-levels, which according to the theory of Bassani et al. may be reasonably long.

The modulation experiments of Chiarotti and Grassano give the most direct evidence for the connection between the L-bands and the F-center [110]: these authors have modulated the ground and excited state population of the F-center at a frequency v by pumping into the F-band. A second weak beam crossing the crystal perpendicularly to the pumping light direction is used to monitor absorption. All absorption originating from the groundstate will be modulated at the frequency v and this modulation of the transmitted light may be detected with a lock-in amplifier. Chiarotti and Grassano have found a modulated signal for each of the transitions commonly attributed to the F-center: the F-, K-, and L-bands. Thus the L-bands originate without doubt from the same groundstate as the F-band.

Once this fact is established, the most interesting problem is that of the lifetime of the resonant electronic state which gives rise to the absorption. The absorption bandwidth does not yield useful information since it is very likely due to the strong electron–phonon coupling, characteristic of the F-center. As we have said, the luminescence yield is about 1%. In the case where the luminescence would originate directly from a relaxation of the L-level into F*, the ionization probability would be about 100 times larger than the typical inverse relaxation time, which in turn is of the order of the optical phonon frequency ($10^{12}\,\mathrm{s}^{-1}$); thus the ionization time could not be longer than $10^{-14}\,\mathrm{s}$. Such a time is about one order of magnitude longer than the optical transition time ($10^{-15}\,\mathrm{s}$). Therefore in this case the L-level lifetime would be long compared to normal scattering levels, in good agreement with theory.

It is thus critical to know the exact origin of the L-excited F-luminescence. An experiment has been performed which yields a reasonably precise answer to this question. The F-emission decay time was precisely measured as a function of temperature upon excitation in the L-bands: this time is found to be identical to that determined by F-band excitation and features its temperature dependence. In order to perform this measurement, Benci and Manfredi [111] have used a pulsed source of weak intensity (in order to avoid effects due to F$^+$ centers. The emission excited in the L-bands of a given F-center originates then from that same center. In fact, if there was sizable recombination of the F$^+$–e$^-$ type, the temperature variation of the luminescence yield would be affected. Although, further experiments are necessary to verify all this, it seems reasonable to assume that the L-excited F-emission is due to the same center which was excited after relaxation. This in turn yields an estimate of the L-level lifetime which confirms well the predictions of the theory of Bassani et al.

4.5. Heavy metal ions in alkali-halides

4.5.1. Optical spectra of Tl-like ions in alkali-halides

The absorption spectra of alkali-halides doped with Tl-like ions (Ga, In, Pb, Sn, Ag, Au) which have groundstate electronic configuration s^2, have been widely studied since the pioneering work of the Göttingen school [112] and the first theoretical interpretation by Seitz [113]; the Seitz model assumes the impurity to enter substitutionally at a cation site, and the experimental evidence for such a hypothesis is now so strong as to leave no alternative possibility.

Following Seitz, the simplest way to describe the impurity electronic states is to assume that they are essentially the same as those of the free ion (see however Bramanti et al. [114] for the role of covalency). Since the lattice environment has not the spherical symmetry of the free ion, it will broaden and possibly split the optical bands of the impurity.

The first excited electronic configuration for the ions we are discuss-

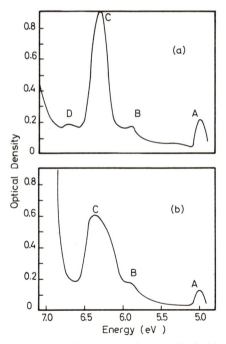

Fig. 4.20. Optical absorption in KCl:Tl:(a), at about −180°C; (b), at room temperature; [after J. E. Eby and K. Teegarden in R. S. Knox, Phys. Rev. 115 (1959) 1095].

ing is sp, which is spherical symmetry is twelvefold degenerate and in a cubic field (O_h symmetry) splits into the following free ion levels: $^3P_0, ^3P_1, ^3P_2, ^1P_1$. Since the ion is not free but rather in a field of O_h symmetry, it is more appropriate to label such levels with the irreducible representations of O_h. Accordingly the levels are: $^3A_{1u}, ^3T_{1u}, ^3E_u + ^3T_{2u}, ^1T_{1u}$. In fig. 4.20, we show the optical absorption spectrum connected with transitions to these levels for the case of KCl:Tl at 77 and 300 K.

The A, B, C, bands correspond to transitions from $^1A_{1g}$ groundstate to the $^3T_{1u}, ^3E_u + ^3T_{2u}, ^1T_{1u}$ states respectively, while the D transition originates from the configuration interaction of the Tl states with the excitonic states of the host crystal [115]: in short, it is due to an excitonic transition localized at the impurity site. All alkali-halides doped with s^2 ions have an absorption spectrum similar to KCl:Tl, but the peak energies of the bands are characteristics of each host crystal dopant combination. The $^1A_{1g} \rightarrow ^3A_{1u}$ transition is not easily observed since it is strictly forbidden. The only dipole and spin allowed transition is $^1A_{1g} \rightarrow ^1T_{1u}$; the $^1A_{1g} \rightarrow ^3T_{1u}$ transition is partially allowed by the singlet–triplet spin–orbit mixing (see the following discussion) while $^1A_{1g} \rightarrow (^3E_u + ^3T_{1u})$ is due to vibronic mixing of ^3E_u and $^3T_{2u}$ with $^3T_{1u}$. This is in good agreement with the experimentally determined increase of the B-band intensity with temperature [116, 117].

The determination of the eigenvalues and eigenfunctions of such heavy ions as Tl^+ (81 electrons) is practically impossible from the point of view of a first principle calculation, even for free ions. It is however easy to obtain the eigenvalues of the sp configuration if the spin–orbit coupling constant ξ_p, the exchange integral G_1, and the King–Van Vleck factor λ [118] are treated as phenomenological quantities to be determined by experiment. In this case the eigenvalues are [119]:

$$E_0 = C - G_1 - \xi_p; \qquad E_1 = C - \frac{\xi_p}{4} - \left[\left(G_1 + \frac{\xi_p}{4}\right)^2 + \frac{1}{2}\lambda^2 \xi_p^2\right]^{1/2}$$

$$E_2 = C - G_1 + \frac{\xi_p}{2}; \qquad E_3 = C - \frac{\xi_p}{4} + \left[\left(G_1 + \frac{\xi_p}{4}\right)^2 + \frac{1}{2}\lambda^2 \xi_p^2\right]^{1/2},$$

$$(4.44)$$

where C is a constant, and λ accounts for the possible difference in the radial wavefunctions of $^1T_{1u}$ and $^3T_{1u}$. The zero energy is referred to the ground-state.

As mentioned, the spin–orbit interaction (which in the heavy ions cannot be treated in the Russell–Saunders scheme) lifts the degeneracy and mixes $|^3T_{1u}\rangle$ and $|^1T_{1u}\rangle$ to give the states:

$$|^1T_{1u}^*\rangle = -v|^3T_{1u}\rangle + |\mu|^1T_{1u}\rangle \qquad (4.45a)$$

$$|{}^3T_{1u}^*\rangle = |v|{}^3T_{1u}\rangle + \mu|{}^1T_{1u}\rangle, \qquad v^2 + \mu^2 = 1 \qquad (4.45b)$$

and this mixing provides the A-band oscillator strength. It can actually be shown that the oscillator strength ratio of the C-band to the A-band f_A/f_C, is given by (Sugano [120]):

$$f_C/f_A = (E_C/E_A)R, \qquad (4.46)$$

where

$$R = \frac{4 - 2x + [6 - 2(2x - 1)^2]^{1/2}}{2 + 2x - [6 - 2(2x - 1)^2]^{1/2}}$$

and

$$x = (E_B - E_A)/(E_C - E_A). \qquad (4.47)$$

A test of Sugano's formula is given for several crystals in fig. 4.21, which shows that there is reasonable agreement. It may be worth mentioning at this point that eqs (4.46) and (4.47) are not, as claimed by several authors, a result due to a molecular orbital approach for the sp configuration; in fact, the ionic and molecular orbital approximations give exactly the same result for f_C/f_A, the expression of which is deducible entirely from group theory [1, 120]. The better agreement of Sugano's formula relative to previous

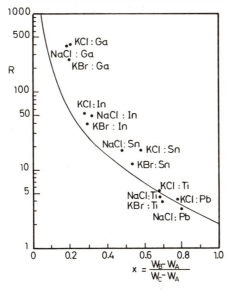

Fig. 4.21. Test of the Sugano formula for several s^2-ions in some alkali-halides [after T. Mabuchi et al. Sci. Light (Tokyo) 15 (1966) 79].

Table 4.7
Energy and G_1 and ξ_p parameters of Tl$^+$ optical transitions in alkali-halides.

	A(eV)	B(eV)	C(eV)	D(eV)	G_1(eV)	ξ(eV)
NaCl:Tl	4.87	5.77	6.19	>7.6	0.29	0.68
KCl:Tl	5.03	5.94	6.36	~7.3	0.28	0.69
					(0.5)[a]	(0.8)[a]
KBr:Tl	4.79	5.58	5.93	6.50	0.23	0.60
KI:Tl	4.40	5.08	5.31	5.52	0.16	0.51
NaI:Tl	4.27	4.95	5.15	5.47	0.14	0.50

[a]Determined theoretically by Bramanti et al. [114].

estimates [121] is due to the fact that Sugano used experimentally determined values for both G and ξ_p, while Williams et al. used the G of the crystal and the ξ_p of the free ion. In table 4.7, we show some of the important parameters connected with Tl$^+$ absorption in some alkali-halides.

As for the B-band, its oscillator strength derives from the mixing of $|^3E_u\rangle$ and $|^3T_{2u}\rangle$ operated by the Jahn–Teller effect: this can be seen by inspection of the whole 12×12 interaction matrix as reported, for instance, by Toyozawa and Inoue [122]. In particular, it is the trigonal modes $\tau_{2g}(Q_4, Q_5, Q_6)$ which give rise to the coupling. It is rather safe to conclude that as far as the peak positions of the absorption bands are concerned the situation is well understood (see e.g. the molecular orbital calculations of Bramanti et al. [114]).

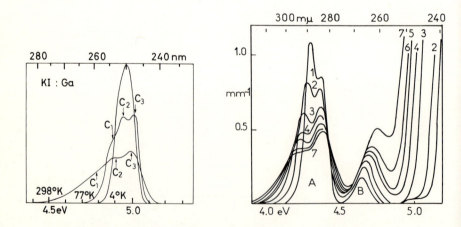

Fig. 4.22. The temperature variation of the A band in KCl:In [122] and KI:Ga:(a), 1:20 K; 2:77 K; 3:133 K; 4:175 K; 5:211 K; 6:243 K; 7:285 K; (b), KI:Ga.

The next step towards the understanding of the absorption mechanism is to explain the triplet or the doublet structure exhibited by some phosphors. In fig. 4.22, we show two typical examples of such structures: the C-band of KI:Ga is a triplet, while the A-band in KCl:In is a doublet. Remembering that the C-band is due to the transition $|^1A_{1g}\rangle \to |^1T_{1u}\rangle$ and that the $|^1T_{1u}\rangle$ state is predominantly spin singlet, this structure is not surprising in view of the discussions of the previous chapter. We shall only add that the temperature dependence of the splitting is very well accounted for by the semiclassical theory of Toyozawa and Inoue [122]. On the other hand the doublet character of the A-band is more intriguing, because it is also due to a A to T transition like the C-band. The difference resides only in the prevalence of spin triplet character in the A-band: and in fact it is the spin–orbit coupling which induces the asymmetry, as discussed by Toyozawa and Inoue[122]: the doublet is thus a very asymmetric triplet, in which two components have coalesced together. Here we shall not dwell on the details but rather give an idea of how the thing works. The band-shape $I(\omega)$ [eq. (3.39)] is symmetric only if the argument of the delta-function

$$\delta(\hbar\omega - E_f^q + E_i^q)$$

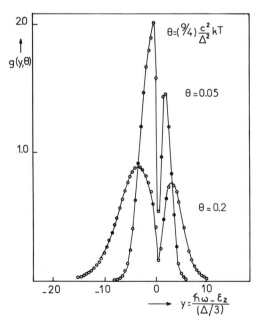

Fig. 4.23. The calculated line shapes of the A-band in the limit of no triplet–singlet mixing $(t \to 0, R \to \infty)$ for various temperatures θ (after Toyozawa and Inoue [122]).

is linear in all the q's. Now, if we consider only the linear JTE and neglect spin–orbit interaction, linearity results from the exact cancellation of the elastic term, which is assumed to be the same for E_f and E_i. But it is relatively easy to see that spin–orbit interaction produces terms quadratic in the q's, other than the elastic energy itself, which do not cancel, thus giving rise to the asymmetry. Since the very existence of the A-band is due to spin–orbit, it is intuitively understood – and can be demonstrated – that its asymmetry is much larger than that of the C-band; in fig. 4.23, we show the A-band calculations of Toyozawa and Inoue for several temperatures in the limit of weak spin–orbit coupling. Another cause of non-linearity in the delta-function argument are, of course, quadratic J–T effects: actually, Honma [123] found that in the temperature behavior of the separate peaks of the C-band the latter effect is predominant over the former. Finally, Cho [124] has calculated C band-shapes by the Monte Carlo method, including the effect of a magnetic field. MCD experiments have confirmed the J–T nature of the C and A band-structures (fig. 4.24) [125].

In conclusion, it seems that the absorption spectra of at least Tl and In are quite well understood in the framework of the semiclassical theory which includes the Jahn–Teller effect. The reason why semiclassical theory works so well is probably due to the fact that the electron–phonon interaction is strong, as can be deduced by the large Stokes shifts for these phosphors. The situation is very different for the Au^- impurity: for instance, in KCl:Au the zero-phonon line may be observed and in this case thus fully quantum-mechanical calculations must be carried out [126].

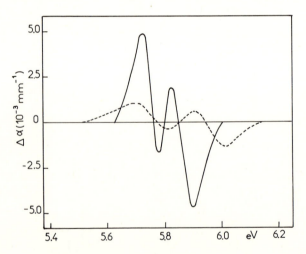

Fig. 4.24. The C-band MCD in KBr:Tl for a field strength of 4.2 kG at 80 K (full line) and 300 K (dashed line) (after Grasso et al. [147]).

4.5.2. *Luminescence properties of Tl-like impurities in alkali-halides*

Alkali-halides crystals (or powders) doped with Tl-like ions feature several emission bands after excitation by ionizing radiation, fast electrons, UV light. The luminescence bands are the same as those observed upon excitation in the A, B, C and D bands; they are broad and structureless. For simplicity we shall discuss only the emission excited in the A-band and refer the interested reader to Fowler [1] for a comprehensive review concerning other excitations.

The emission properties are in general more complicated than for absorption. The monovalent impurity phosphors may be grouped into three categories: the first two show two emission bands, called the A_T (high energy) and A_X (low energy), while the third category has apparently only one emission band (see table 4.8). Groups I and II and differentiated by the relative temperature dependence of A_T and A_X emissions. In group I, at low temperature only the A_T band is observed, whereas the A_X emission appears and increases as temperature increases: the total emitted intensity remains approximately constant, and above about 60 K only the A_X band is observed. The temperature behaviour of group II phosphors (KBr:In and KBr:Tl for instance is more complicated in the high temperature region (> 150 K for KBr:Tl). We shall discuss in some detail the group I phosphors. The main problems are: why are there two emission bands and what is the reason for the observed temperature dependence. There are however other significant and puzzling experimental features. For instance, the A_T emission is linearly polarized (upon excitation with linearly polarized light)

Table 4.8
Classification of s^2 heavy ions doped alkali-halides from the characteristics of their emission (from Fukuda [131]).

	KI:Ga	KBr:Ga	KCl:Ga	
A_X(eV)	2.04	2.24	2.35	
A_T(eV)	2.47	2.74	2.85	Group I
	NaCl:Ga	KI:In	KI:Tl	
A_X(eV)	2.45	2.20	2.89	
A_T(eV)	3.10	2.81	3.70	
	KBr:In	KBr:Tl		
A_X(eV)	2.46	3.50		Group II
A_T(eV)	2.94	4.02		
	KCl:In	NaCl:In	KCl:Tl	
A_X(eV)	—	—	—	Group III
A_T(eV)	2.95	3.05	4.17	

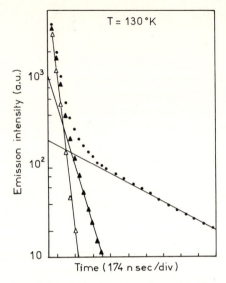

Fig. 4.25. Time behavior of the A_X luminescence for a KI:Tl crystal excited in the A-absorption band. From this spectrum three different slopes, each one characterized by a different value of the lifetime, can be determined (after Benci et al. [130]).

along a tetragonal [001] direction, while A_x is either non-polarized or shows a weak trigonal [111] polarization [127, 128, 129]. Furthermore, the decay time of the A_X emission is complex; an example is given in fig. 4.25, referring to KI:Tl [130]. At 130 K the A_X emission decays with three components, of 25, 100 and 1000 ns, respectively, the lower decay time being strongly temperature-dependent [130].

Further complications arise when the effects of electric, magnetic and stress fields are considered. We shall discuss them in some detail later, in the framework of a model which accounts for the intrinsic properties of the emission.

If we assume that the emissions take place from the minima of the potential energy surface in the excited $^3T_{1u}$ state, a mechanism must be found which justifies the existence of two kinds of minima at different energies. Furthermore, one of these minima must have tetragonal symmetry since the emission is polarized along a [100] direction; the other kind should be trigonal in the phosphors where the A_X emission is [111] polarized, while no symmetry restrictions are imposed on those phosphors where the emission has not been observed to be polarized. From the theoretical results presented in the previous chapter, we know that the linear JTE alone can be of little help since it produces only one kind of minimum (either tetragonal or trigonal) and can never stabilize orthorhombic distortions. There are however two mechanisms which – in addition to the linear JTE – can provide the required coexistence of minima of different energies on the lowest energy surface of the relaxed excited state: one is the spin–orbit mixing between $^1T_{1u}$ and $^3T_{1u}$, the other is the quadratic JTE and/or anharmonicity in the restoring forces. It would be out of place here to enter into the mathematical details of these effects; we shall try rather to give a qualitative description of the way in which they operate and to discuss to what extent can the resulting models explain the phenomenology of phosphors.

For the heaviest ions especially, the spin–orbit interaction is so strong that it cannot be treated, with any reasonable degree of accuracy, within the Russell–Saunders scheme, where the singlet–triplet mixing is neglected. The effect of the interaction is depicted, very schematically, in fig. 4.26, where Q is a generic J–T active coordinate. The two uncoupled parabolas of fig. 4.26(a), schematize the pure $^1T_{1u}$ and $^3T_{1u}$ states, their minima being displaced from the origin by the JTE; in this approximation the two energy curves are allowed to cross. Fig. 4.26(b) shows, on the other hand, what happens when the spin–orbit mixing is taken into account: the mixed states now repel, and if such repulsion is strong enough new minimum T* appear on the lowest state $^3T_{1u}^*$. This basic mechanism has been studied in detail by Ranfagni [132] and Ranfagni and Viliani [133] who discussed J–T coupling only to the tetragonal coordinates Q_2, Q_3. A typical cross-section is shown in fig. 4.27, and shows that the $^3A_{1u}$ state in the neighborhood of the X minimum is very close in energy to the $^3T_{1u}^*$ state; moreover, the energy curve on which the X minima lie is doubly degenerate.

These results lead to the following model for the emission: A_T originates from T* minima, A_X from X minima; the fast component of A_X is caused by the spin–orbit allowed $^3T_{1u}^* \to {}^1A_{1u}$ transition, while the slow component comes from the (possibly) hyperfine interaction allowed [134] $^3A_{1u} \to {}^1A_{1g}$ transition. As for the intermediate lifetime, we shall discuss its origin later. The degeneracy of $^3T_{1u}^*$ near the X minimum accounts for the magnetic circular polarization of the A_X emission [135]. This model is particularly well suited for Tl phosphors, where the A_T emission has a single lifetime (thus requiring no trap level under the T* minima) and weak circular polarization under magnetic field (and thus no degeneracy of T* minima); however it is not versatile enough to interpret results for Ga and In; in this case in fact the results of Fukuda et al. [136] indicate that both minima are degenerate and have the trap level underneath. Such a level scheme may be obtained if quadratic JTE and/or anharmonic restoring terms are included in the potential: these would allow the coexistence of different kinds of minima on the potential surface [137]. In fig. 4.28, we present a resumé of the possible arrangements which may arise from higher order vibronic effects.

Scheme 1 in fig. 4.28, refers to spin–orbit caused coexistence. Thus, the model which emerges from this picture (and which is adequate for phosphors

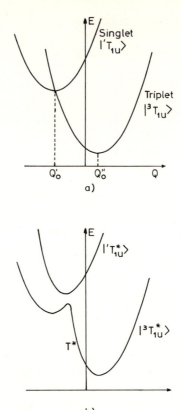

Fig. 4.26. Schematic diagrams of ^1T_u and ^3T_u APES sections along coordinate Q: (a), no spin–orbit interaction; (b), spin–orbit interaction included.

exhibiting either MCD on both emissions or trigonal polarization of A_X) is the following: A_T originates from eT (i.e. elongated tetragonal) minima, A_X from O (orthorhombic) minima. Scheme 3, consisting of tetragonal and trigonal minima, might also explain the experimental features if we assume that the unpolarized A_X emission originates from slightly distorted trigonal minima which do not preserve polarization; apart from coexistence, the important features of schemes 2 and 3 are the degeneracy (near degeneracy for trigonal minima) at the minima, and the presence of an underlying trap level. From this necessarily short discussion, it emerges that the J–T model for phosphor emission (including quadratic JTE and spin–orbit mixing) can account for most of the experimental aspects of such systems. In this sense, the alternative model, repeatedly advanced, and which assigns one of the emissions to a perturbed excitonic state localized at the impurity site (see

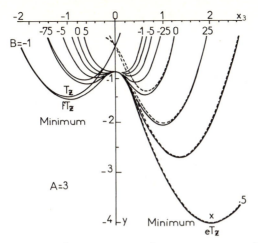

Fig. 4.27. Cross sections of the $^3T_{1u,z}$ (solid lines), $^3T_{1u,x(y)}$ (dashed lines) $^3A_{1u}$ APES's, computed for different values of the parameter $B = 3^{1/2}\rho b_{\varepsilon\varepsilon}/b^2$ between -1 and 0.5, and for $A = 12\rho(1 + a_{\varepsilon\varepsilon})b^2$. There are two stationary points, T_z and X on the $^3T_{1u}$ cross section and only the deeper one is minimum. Similar, the $^3A_{1u}$ cross section shows the fT_z and the eT_z stationary points, and the deeper of them is a minimum (from Ranfagni and Viliani [133]).

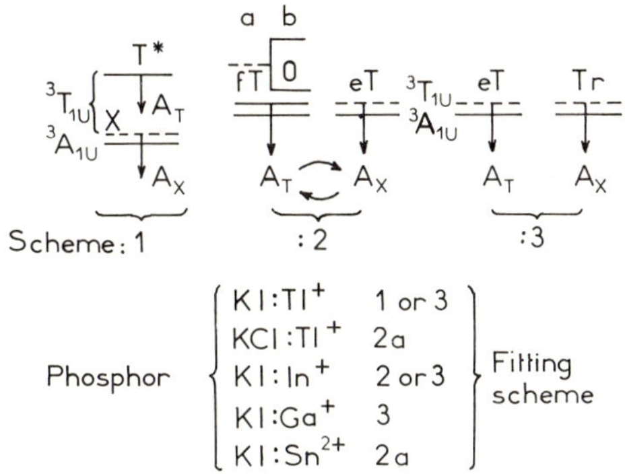

Fig. 4.28. Energy levels schemes and tentative band assignment of the double A-emission of KI:Tl-type phosphors (dashed levels are doubly degenerate). Scheme 1 is obtained in the linear J–T framework; schemes 2 and 3 are obtained by including quadratic terms. Scheme 2a consists of non-coexisting fT and eT tetragonal minima; inclusion of trigonal coordinates shows that fT minima are actually of orthorhombic symmetry (O) and can coexist with eT tetragonal minima; therefore, the band assignment must be reversed (scheme 2b). Scheme 3 consists of coexisting tetragonal and trigonal minima (after Bacci et al. [141]).

for instance Donahue and Teegargen [138] and refs. therein) is left little breathing room. Even though this model (or models) might in principle fit some of the data, it is not of sufficiently general applicability and requires *ad hoc* assumptions thus appearing clearly inadequate relative to the J–T model.

Before we pass on to consider non-radiative processes in the excited state of Tl-like ions (important for the understanding of the T behavior of the A_X and A_T emissions and their complex decay schemes) it may be worth discussing briefly the effects of external fields on luminescence. The effect of a magnetic field is reasonably well understood and fits more or less accurately into the J–T scheme, as already mentioned. Recently, EPR has been detected on the relaxed excited state of KBr:Ga [139]: the EPR spectrum of A_T with H parallel to [001] is qualitatively analogous to the A_X spectrum with H parallel to [011], and this is confirmation of Trinkler and Zoloukina's previous discovery of the coexistence of trigonal and tetragonal minima (129). Some interesting indications come from hydrostatic pressure experiments [140]: on the one hand, the experiments are well fitted within the spin–orbit scheme for Tl phosphors in KCl-type host crystals (sixfold coordination); on the other hand, for CsCl-type phosphors (eightfold coordination) there are three A-band excited emissions. Drotning and Drickamer [140] argue that three different kinds of minima exist on the $^3T_{1u}$ surface, and note that such an occurrence is not excluded by the theo-

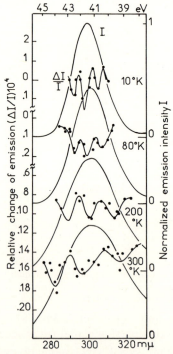

Fig. 4.29. Relative change of KCl:Tl emission excited in the A-band, as a result of an external electric field versus energy at various temperatures. The corresponding emission bands are also reported (after Ranfagni et al. [143]).

retical analysis of Bacci et al. [141]. Electric field effects in emission are more difficult to interpret. For instance, a triplet structure has been detected in the modulated emission of KCl:Tl and KBr:Tl under AC electric field [142]. Ranfagni et al. [143] measured the temperature dependence of the triplet structure in the 3000 Å emission of KCl; Tl (such a structure is shown in fig. 4.29, and the splitting was found to vary as T). These authors interpreted such splitting by analogy with the structures observed in absorption (in zero field), but were fully aware of the difficulties of such an interpretation. Briefly, it is difficult to understand why the upper component of the triplet should be populated at 10 K, if the components of the triplet correspond to the three J–T split electronic components of $^3T_{1u}$. Furthermore, other authors find no evidence of electric field induced structures in Tl emission†.

4.5.3. Non-radiative processes in the excited state of Tl

As we have mentioned in the previous section, A_X and A_T intensities are temperature dependent. In KI:Tl for instance (fig. 4.30), A_T is strongly

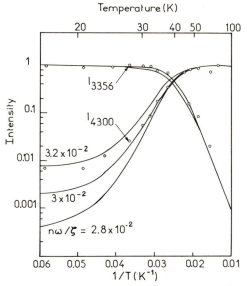

Fig. 4.30. Temperature dependence of the intensities of A_T (3356 Å) and A_X (4300 Å) emissions in KI:Tl. Circles: experimental data of Illingworth [148]; full lines: after eqs. (4.49), (4.50) with $A = 3$, $g = 0.4$, $\tau = 2.5 \times 10^{-8}$s, $\hbar\omega/\rho = (2.8 - 3.2) \times 10^{-2}$ (after Ranfagni et al. [149]).

†U. M. Grassano, Private Communication.

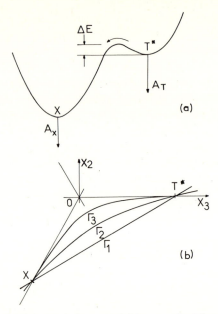

Fig. 4.31. Typical potential shape (a) along a path (b) connecting T* and X minima from which the A_T and A_X emission arise (after Ranfagni et al. [149]).

prominent at low temperature, and then feeds A_X as T is increased. This can be interpreted quantitatively within the spin–orbit model, by calculating the rate of non-radiative transitions from the higher T minimum to the lower X minimum [144]. The calculation may be carried out using the WKB method which greatly simplifies the procedure (for a justification of this procedure in multi-dimensional spaces see also Ranfagni et al. [144]), assuming the system to move from one minimum to the other along the classical minimum action trajectory. In the q_2, Q_3 space, the situation is depicted in fig. 4.31; we must choose the actual path according to:

$$\delta \int_\Gamma [2\mu(E - V)]^{1/2} \, ds = 0, \tag{4.48}$$

where μ is the effective mass in the normal coordinate space, E the energy, V the potential energy, and the integrals are computed along the paths Γ_i connecting the X and T* minima. Once Γ is determined numerically (the potential shapes are known – ref. [133]) for each vibrational energy level within T*, the total radiationless transition rate $W_{T \to X}$ as a function of temperature is given by:

$$W_{T \to X}(T) = \left[1 - \exp\left(-\frac{\hbar\omega}{kT}\right)\right] \sum_{n=0}^{\infty} \nu D_n \exp\left(-\frac{n\hbar\omega}{kT}\right) \tag{4.49}$$

where

$$D_n = \left[1 + \exp\left(\frac{2}{\hbar}\int_a^b |p|\,dx\right)\right]^{-1},$$

is the transmission coefficient for the nth vibrational state (Landau and Lifshitz [145]) within the T* well, $v = 2\pi\omega$ is the vibrational frequency, $\hbar\omega$ is the energy separation between adjacent vibrational levels and $p = [2\mu(E - V)]^{1/2}$ plays the role of momentum in normal coordinate space. Using reasonable approximations, the intensities of A_X and A_T emissions as a function of temperature may be evaluated:

$$I_{A_T}(t) = [1 + 2\tau W_{T\to X}(T)]^{-1}$$

$$(4.50)$$

$$I_{A_X}(T) = 1 - I_{A_T}(T)$$

where τ is the lifetime for the $^3T_{1u} \to {}^1A_{1g}$ transition from T*. These quantities depend more or less explicitly on the parameters which characterize the system: these enter D_n through the potential V. The full lines of fig. 4.30 have been computed using the values specified in the caption, which are reasonable for KI:Tl: the agreement seems to be rather satisfactory, and even the larger discrepancy at higher temperature may be understood if the diminution of level spacing with increasing vibrational energy is included in the calculation.

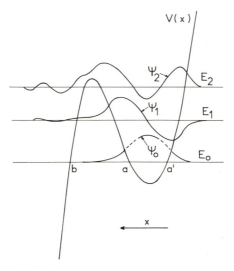

Fig. 4.32. Eigenvalues and eigenfunctions within the well T* obtained by the WKB procedure showing tunnelling through the energy barrier (after Ranfagni et al. [149]).

The same model can be used to explain the existence of three different lifetimes in the KI:Tl-type phosphor luminescence (A-emission) [130, 146]. To understand the mechanism qualitatively, let us refer to fig. 4.32; we see that different vibrational states have quite different tunnelling probabilities, and hence different lifetimes in well T^*. The inverse lifetime of the A_X emission is the sum of the inverse lifetimes of the radiative $^3T_{1u} \rightarrow$ $^1A_{1g}$ and the tunnelling $T^* \rightarrow X$ transitions. The lowest level turns out to have so small a tunnelling probability that, practically, it does not contribute to the A_X intensity; the higher levels have short tunnelling times and the corresponding emission lifetime is limited by the radiatiye lifetime; they contribute the fast component of A_X; finally, the E_1 level has a tunnelling time of the order of 100 ns, and this explains well the third component of the A_X emission. The agreement with experiment is thus good as far as lifetimes are concerned, but it becomes worse when intensities are considered. Such a partial failure of the model indicates that the model itself is over-simplified and that a full account of the radiationless processes must involve more complex mechanisms. For instance, the calculation should be extended to the full six-dimensional configuration coordinate space and not just be limited to the tetragonal coordinates. The calculations in six dimensions are however difficult.

In conclusion, it is apparent that in recent years a much more quantitative understanding of Tl emission has been achieved, and certainly the qualitative aspects of the problem are now settled. In particular, the role of the J–T effect, in general, and particularly the quadratic J–T in producing the double emission structure is now not in doubt, as also shown in the experiments of Trinkler and Zoloukina [129].

4.6. Non-radiative transitions: an example, the triplet state of F_2-centers in alkali-halides

We have largely quoted in chapter 3 the paper by Huang and Rhys [150] describing the effect of electron–phonon coupling on the shape of absorption and emission bands. The second part of their paper is little known where they discuss how the same coupling can induce a non-radiative transition between the F-center relaxed excited state and its groundstate and how this non-radiative transition can compete with the radiative one. To present the theory of non-radiative transitions, let us just discuss the process of detrapping an electron from an acceptor in an insulator. In the trap, the electron is submitted to a potential which presents fluctuations by phonons. If the electron is following these fluctuations adiabatically, it is always in a steady state and cannot escape from the trap; but if it cannot follow these fluctuations, it could escape to the conduction band. Let us present a very simple classic analogy: if I put water in a plastic bag and deform it slowly, all water molecules are following adiabatically the deformation and the surface remains well defined. If now, I deform the bag more and more quickly, drops are jumping higher and higher above the mean surface because water molecules cannot follow the bag deformation adiabatically anymore. In this part of this chapter, we shall use the F_2-centers as a good example of non-radiative transition and more precisely non-radiative transitions between the triplet groundstate and the singlet groundstate. In the first part, we shall present experimental evidence for the triplet state of F_2-centers and for a competition between radiative and non-radiative decay of this state. Then, in the second part the theory of non-radiative transitions is presented and in the last part, it is shown that this theory explains well the properties of the F_2-center triplet state.

4.6.1. Experimental results on the metastable triplet state of F_2-centers [151]

Before the nature of F-aggregate centers was known, it was observed that some optical transient absorption bands were related to them. For example in an alkali-halide crystal containing F- and F_2-centers illuminated at low temperature with F light a decrease of the F_2-absorption band and the appearance of new transient absorption bands are observed. After turning off the excitation light (pumping light) these transient bands decrease with characteristic decay time (100 s and 12 s for KCl and KBr respectively at 4 K [152]). The metastable state related to these absorption bands has been identified by Seidel [153] as being the triplet state of F_2-centers which is obviously paramagnetic. Transient absorption bands correspond to transitions from the triplet groundstate $^3\Sigma_u$ to higher excited states (fig. 4.33).

The spectroscopy of this triplet state is now well known [154, 155]. Particularly, the triplet–singlet conversion is enhanced if the triplet system is in an excited state as it has been observed by Haarer and Pick [154]; then it is possible to use this fact to determine precisely the symmetry of each level [155] by exciting the triplet states with polarized light which are represented in fig. 4.33. But the most interesting phenomenon is the triplet–singlet transition from the triplet groundstate. It has been observed by Farge [156] that the lifetime of this state measured at 4 K in KCl via the 685 nm triplet–triplet transient absorption band depends strongly upon the degree of aggregation of F-centers in an additively colored crystal. This can be explained by assuming that an interaction between an F_2- and a neighbouring F-center can reduce the lifetime. In the following we shall discuss only the case of fully isolated F_2-centers giving a triplet lifetime

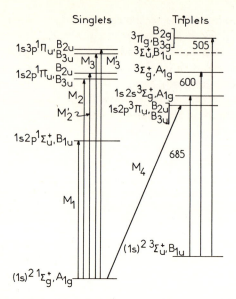

Fig. 4.33. Energy levels of singlet and triplet states of F_2-centers (after Engström [155]).

around 100 s at 4 K when it can be as short as 60 s when they are interacting with F-centers.

The triplet lifetime τ decreases when the temperature increases and its variation has been followed up to 250 K in KCl by measuring the triplet concentration via the linear dichroism induced at 685 nm ($^3\Sigma_u \rightarrow {}^3\Sigma_g$ transition) with a crystal pumped with polarized F light [155, 157].

A radiative and a non-radiative process are competing for the decay of the triplet state. The radiative one is temperature independent giving a radiative probability $1/\tau_R$ when the non-radiative one $1/\tau_{NR}$ varies with temperature:

$$\tau^{-1} = \tau_R^{-1} + \tau_{NR}^{-1}. \tag{4.51}$$

values measured by Ortega in KCl and KBr are listed on table 4.9.

The radiative process has been confirmed by the observation of an

Table 4.9
Radiative and non-radiative lifetimes of the triplet state of F_2-centers in KCl and KBr at 4 K [151].

	KCl	KBr
τ_R	105 s	12.5 s
τ_{NR}	600 s	200 s

extremely weak fluorescence in the infrared corresponding to the $^3\Sigma_u \rightarrow$ $^1\Sigma_g$ transition and having the typical triplet lifetime [158]. Ortega [159] has measured the spectral distribution of this emission with a technique which is discussed in his paper. Fig. 4.34, shows this emission spectrum in KCl, KBr and KI, the resolution is not very good, but a zero-phonon line can be observed and a phonon structure which gives E_{TS}, the energy difference between the triplet and the singlet groundstates and the coupling of the singlet groundstate with phonons, i.e. the relaxation between the two states or, in the linear coupling approximation (see § 3.2.2) the Huang–Rhys parameter S with main phonons. The determination of E_{TS} and S was important to interpret the non-radiative process as it will be shown later.

The non-radiative probability varies with temperature according to the following expression, up to 120 K in KCl and 100 K in KBr:

$$1/\tau_{NR} = (1/\tau_{NR}(0)) \coth(\hbar\omega/2kT). \tag{4.52}$$

The additional probability of the triplet–singlet transition induced by a neighboring F-center has the same temperature variation than $1/\tau_{NR}$ of isolated defects and Ortega concluded that this interaction increases the non-radiative probability.

The system is now sufficiently well defined to perform theoretical interpretations. Position of singlet and triplet levels are known as well as their symmetry and good wavefunctions can be built using well known F-center wavefunctions measured by ENDOR and the Heitler–London method.

4.6.2. Interpretation of the radiative process [157, 161]

It is well known that a radiative process in such a system can occur only if a spin–orbit interaction mixes singlet and triplet states [160]. Then real singlet and triplet groundstates have to be written as follow:

$$|S'_0\rangle = |S_0\rangle + \sum_{i,k} \frac{\langle S_0| \mathcal{H}_{So}| T_i^k\rangle}{E_{ST} - E_{T_1}^k} |T_i^k\rangle$$

$$\tag{4.53}$$

$$|T_1^{k'}\rangle = |T_1^k\rangle + \sum_j \frac{\langle T_1^k| \mathcal{H}_{So}| S_j\rangle}{E_{T_1}^k - E_{S_j}} |S_j\rangle$$

where: $\mathcal{H}_{So}$ is the spin–orbit operator; $|S_j\rangle$, the wavefunction of the jth singlet state; $|T_i^k\rangle$, the wavefunction of the ith triplet state and the kth component of the spin multiplet.

The radiative transition probability is proportional to the square of the

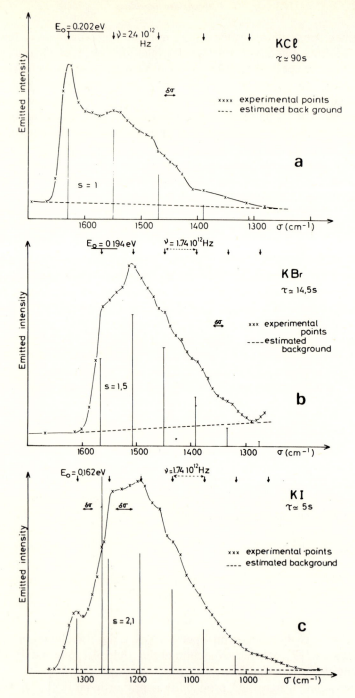

Fig. 4.34. Emission spectra corresponding to the radiative transition from the triplet to the singlet groundstate (after Ortega et al. [158]).

matrix element $\mathbf{M}$:

$$\mathbf{M} = \langle S'_0 | e\mathbf{r} | T_1'^k \rangle. \qquad (4.54)$$

Without discussing the theoretical analysis in detail, let us just focus on two main points; what are the sublevels of the triplet groundstate which can decay by a radiative process and what is the origin of the spin–orbit interaction. The first point is quite simple to explain. F_2-centers have D_{2h} symmetry; the singlet groundstate transforms as A_{1g}, the triplet groundstate sublevels transforms like B_{1u}, A_u and B_{3u} when r, in eq. (4.54), transforms like B_{1u}, B_{2u}, and B_{3u} when it is respectively parallel to x, y, and z, y being the axis of the defect. From these simple considerations, it can be seen that the A_u level cannot emit when B_{1u} and B_{3u} will emit light polarized along x and z (x parallel to [110] and z parallel to [001] for a defect along [110]). This situation is similar to the self trapped exciton case (Fig. 4.19).

In principle, for a molecule having the D_{2h} symmetry, the spin–orbit operator cannot mix Σ and Π states because there is no degenerate state and the lifetime of the triplet state should be infinite. In the case of F_2-centers, the spin–orbit operator can induce such a mixing because F_2-center wavefunctions contain a small amount of p-like atomic wavefunctions on neighboring ions which comes from the orthogonalization of the F-center electron wavefunction with these ions (fig 4.35). For the F-center:

$$\Phi'_A = N[\Phi_A - \sum_\alpha S_\alpha \Phi^\alpha] \qquad (4.55)$$

with

$$N = (1 - \sum_\alpha S_\alpha^2)^{1/2} \quad \text{and} \quad S_\alpha = \langle \Phi_A | \Phi^\alpha \rangle.$$

By using F-center wavefunctions with mixing coefficients S_α determined by ENDOR and building F_2-center wavefunctions with the Heitler–London method, it is possible to calculate exactly the radiative transition probability [161]. Results are shown in table 4.9 and are compared with experimental results; the good agreement is a confirmation of the correctness of the theory.

Table 4.10
Radiative transition probabilities of the triplet state of F_2-centers in several alkali-halides experiment (4 K) and theory. From Ortega [157].

	KCl	KBr	KI	NaCl	LiF
$1/\tau_R$ exp.	1.1×10^{-2}	8×10^{-2}	15×10^{-2}	0.5×10^{-3}	—
$1/\tau_R$ theor.	1.2×10^{-2}	3.5×10^{-2}	7×10^{-2}	2.5×10^{-3}	3×10^{-3}

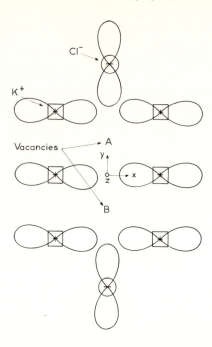

Fig. 4.35. F_2-centers in alkali-halides and p-like contributions to $\sum$ states on neighboring ions of the two vacancies. This small contribution to $^1\sum_g$ and $^3\sum_u$ allows the very small spin–orbit mixing between these two states [161].

4.6.3. *The non-radiative mechanism* [161]

Let us now discuss the non-radiative process which has been observed experimentally in the decay of the singlet–triplet state of F_2 and more precisely its variation with temperature. Theory of non-radiative transitions has been extensively studied and a good review paper has been written by Jortner et al. [162] for molecules. In our case, F_2-center can be considered as a small molecule with a large number of degrees of freedom (all the modes of the crystal!). Then, what is the probability for the system to evolve from the triplet groundstate to a high vibronic level of the singlet groundstate as shown in fig. 4.36. It is clear from this classical picture that only an "odd" phonon can induce this transition because the two states have opposite parity; let q be this mode and $\hbar\omega_q$ the associate energy; this mode is usually called "promoting mode". Then, the final state will have an energy differ-

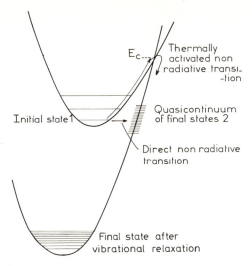

Fig. 4.36. Schematic representation of a direct non-radiative transition and a thermally activated non-radiative transition.

ence $\pm\hbar\omega_q$ with the initial state. Therefore, it is clear that a Franck–Condon integral will play an important role; this integral will represent the vibrational overlap between the initial state and the final state. Such a Franck–Condon overlap varies critically with the vibrational relaxation of the triplet state towards the singlet state which comes from even vibrational modes which are usually called "accepting modes". These modes allow also the system to relax from the "final" state to the bottom of the singlet co-ordinate configuration curve.

We consider that the system has been prepared in the triplet groundstate and we describe all the states using the Born–Oppenheimer approximation (BO) in which electronic states depend parametrically upon nuclear co-ordinates. But a BO state is a stationary state as we said in the introduction of this section which means that it would remain unchanged forever. In fact, this state is mixed by non-BO terms of the Hamiltonian which act like a time dependent perturbation (applied at $t = 0$); this perturbation can thus induce a transition between the two states.

The BO states can be written:

$$\Psi_{n,m}(\mathbf{r}, \mathbf{Q}) = \Phi_n(\mathbf{r}, \mathbf{Q})\, \chi_m(\mathbf{Q}), \qquad (4.56)$$

where: $\mathbf{r}$ represents electron coordinates; $\mathbf{Q}$, represents coordinates of normal modes; Φ_n, represents the electronic wavefunction for the electronic state "n" which depends parametrically upon nuclear coordinates; χ_m, represents the vibrational wavefunction describing the "m" state of vibrational modes.

Let $\mathcal{H}$, the total Hamiltonian of the system, the wavefunction $\Psi_{n,m}$ is an eigenfunction of the BO Hamiltonian $\mathcal{H}_{BO}$ such that:

$$\mathcal{H} = \mathcal{H}_{BO} + \mathcal{H}', \tag{4.57}$$

where $\mathcal{H}'$ is the non BO part of the total Hamiltonian. Then, it can be easily shown that:

$$\mathcal{H}'\Psi_{n,m}(\mathbf{r}, \mathbf{Q}) = T\Phi_n(\mathbf{r}, \mathbf{Q})\,\chi_m(\mathbf{Q}) - \Phi_n(\mathbf{r}, \mathbf{Q})\,T\,\chi_m(\mathbf{Q}) \tag{4.58}$$

where [161]

$$T = \sum_i \frac{\hbar^2}{2M_i} \frac{\partial^2}{\partial Q_i^2} \tag{4.59}$$

the summation being carried out over all the modes of the system. Let us consider now the transition from the initial state "1" to a final state "2". $\mathcal{H}'$ mixes the state "1" with the quasi-continuum of state "2". Bixon and Jortner [163] have shown that the golden rule is valid if the energy associated with $\mathcal{H}'$ is very much larger than the distance between two vibronic levels of the final state. This is certainly the case of F_2-centers because of the very large number of modes. Then:

$$\tau_{1\to2}^{-1} = (2\pi/\hbar)\rho_2|\langle\Psi_1|\,\mathcal{H}'\,|\Psi_2\rangle|^2, \tag{4.60}$$

where ρ_2 is the density of states of the final quasi-continuum. But several initial states have to be considered and their population depends upon the temperature. Therefore, it is necessary to make a thermal average, and:

$$\tau_{1\to2}^{-1} = (2\pi/\hbar)\,\{\rho_2|\langle\Psi_1|\,\mathcal{H}'\,|\Psi_2\rangle|^2\}_{\text{Th Av}} \tag{4.61}$$

4.6.3.1. Calculation of matrix elements. Generally, two situations have to be considered. In the first, initial and final states have the same spin multiplicity and this case is called "internal conversion". To calculate the transition probability, one expands electronic wavefunctions to the first-order with normal coordinates:

$$\Phi_n(\mathbf{r}, \mathbf{Q}) = \Phi_n(\mathbf{r}, 0) + \sum_i \left(\frac{\partial\Phi_n(\mathbf{r}, \mathbf{Q})}{\partial Q_i}\right)_{Q=0} Q_i \tag{4.62}$$

then, using the preceding equations, it is possible to calculate the transition probability. In the second situation, which is the case of the triplet singlet transition of the F_2-centers, the spin multiplicity of initial and final states is different; this case is called "intersystem crossing". Here, a transition is possible only if the spin–orbit interaction mixes states of different multiplicity but this interaction varies also with nuclear coordinates and has to be expanded:

$$\mathcal{H}_{SO}(\mathbf{Q}) = \mathcal{H}_{SO}(0) + \sum_i \left(\frac{\partial \mathcal{H}_{SO}(\mathbf{Q})}{\partial Q_i} \right)_{Q=0} Q_i. \qquad (4.63)$$

Then, the electronic wavefunction of the state n, mixed by $\mathcal{H}_{SO}$ with state p, can be written:

$$\Phi'_n(\mathbf{r}, \mathbf{Q}) = \Phi_n(\mathbf{r}, \mathbf{Q}) + \sum_{p \neq n} \frac{\langle \Phi_p(\mathbf{r}, \mathbf{Q}) | \mathcal{H}_{SO}(\mathbf{Q}) | \Phi_n(\mathbf{r}, \mathbf{Q}) \rangle}{E_p - E_n} \Phi_p(\mathbf{r}, \mathbf{Q}), \qquad (4.64)$$

in which Φ_n and $\mathcal{H}_{SO}$ are expressed by eqs. (4.62), and (4.63). The total wavefunction of the state n, in the vibrational state m, can be written:

$$\Psi'_{n,m}(\mathbf{r}, \mathbf{Q}) = \Phi'_n(\mathbf{r}, \mathbf{Q}) \chi_m(\mathbf{Q}). \qquad (4.65)$$

Formally, the transition probability can now be calculated using eqs. (4.58), (4.59), (4.61) and (4.65). In the next section, we apply this theory to calculate the non-radiative transition probability between the triplet ground-state $^3\Sigma_u$ and the singlet groundstate $^1\Sigma_g$ of the F_2-center.

4.6.3.2. Application to the triplet–singlet transition of the F_2-centers. It would take too long to go into the calculations of Farge et al. [161] so we shall summarize their results. They assumed that promoting and accepting modes have the same frequency and that the static spin–orbit interaction mixes the initial state ($^3B_{1u}$ or $^3B_{3u}$) with only one singlet excited state ($^1B_{1u}$ or $^1B_{3u}$); this last approximation is justified by the good agreement between theory and experiment for the radiative process where it has been carried out. They finally obtained:

$$\tau_{NR}^{-1} = (\pi\hbar/M)|N_1 + N_2|^2 (n'_q + 1) R(E_{TS} + \hbar\omega_q, T), \qquad (4.66)$$

where M, is the mass associated with the normal mode ω_q; n_q, is the occupation number of the same mode. N_1, corresponds to the following process: the initial state is already mixed with a singlet of the same symmetry by the static spin–orbit interaction; the non-BO part of the Hamiltonian induces a transition which is allowed by this small singlet contribution to the initial state. N_2, corresponds to the following process: the dynamical part of the spin–orbit interaction can induce non-adiabatic transitions between the pure triplet initial state and the pure singlet final state. $R(E, T)$ is the classical Frank–Condon integral which has been introduced in chapter 3.

Let us first discuss the temperature variation of $1/\tau_{NR}$ in eq. (4.66). $(n'_q + 1)$ varies like $\coth(\hbar\omega_q/2kT)$; this behavior is effectively observed at low temperature. At higher temperature the Franck–Condon integral varies rapidly with T and screens the slow variation of $(n_q^l + 1)$ as it is observed up to 180 K in KCl. However above this temperature, the non-radiative transition probability increases more rapidly than expected from eq. (4.66). This high temperature behavior can be easily understood by a classical process; triplet and singlet configuration coordinate curves are crossing each other at an energy E_c above the triplet groundstate (fig. 4.36). Then, a triplet–singlet transition can occur if the system is thermally activated to this energy; within this model, an activation energy is obtained which is of the right order of magnitude compared with the one which can be calculated with the configuration coordinate diagram obtained from fluorescence measurements on the radiative transition.

Farge et al. [161] have calculated N_1 and N_2 for KCl and KBr. They found that they are comparable in KBr and that in KCl, $N_1 \simeq 0.3 N_2$. This result is interesting because it shows that the dynamical part of $\mathcal{H}_{so}$ can play an important role for non-radiative transitions between triplet and singlet states. As far as we know, this point is still under discussion for molecules.

To conclude this section, it is necessary to say that a calculation of the non-radiative transition probability at 0 K is extremely difficult. In eq. (4.66), calculations of N_1 and N_2 are expected to yield quite good results; the same wavefunctions are used as in the radiative case which gave very good results. The uncertainty comes from the evaluation of the Franck–Condon integral between the initial state with a low vibrational quantum number and the final state with a very high vibronic excitation; this integral varies extremely rapidly with the energy and a very small error in the configuration coordinate diagram can give a change of several orders of magnitude for this quantity.

References to chapter 4

[1] W. B. Fowler, in: Physics of Color Centers, ed. W. B. Fowler, (Academic Press, New York, 1968).

[2] B. S. Gourary and F. J. Adrian, Solid St. Phys. **10** (1960) 127.

[3] J. J. Markham, in: F. Centers in Alkali-Halides (Academic Press, New York, 1966).

[4] A. M. Stoneham, Theory of Defects in Solids (Clarendon Press, Oxford, 1975).

[5] M. Cardona, in: Modulation Spectroscopy (Academic Press, New York, 1970).

[6] J. H. Van Vleck, Phys. Rev. **74** (1948) 1168.

[7] C. H. Henry, S. E. Schnatterly and C. P. Slichter, Phys. Rev. **137A** (1965) 583.

[8] R. Romestain and J. Margerie, Compt. Rendus (Paris) **258** (1964) 2525; J. Gareyte and Y. Merle d'Aubigne, Compt. Rendus (Paris) **258** (1964) 6393; N. V. Karlov, J. Margerie and Y. Merle d'Aubigne, J. Phys. (Paris) **24** (1963) 717.

[9] F. Lüty and J. Mort, Phys. Rev. Letters **12** (1964) 45; J. Mort, F. Lüty and F. C. Brown, Phys. Rev. **137A** (1965) 566.

[10] D. Y. Smith, in: Int. Symp. on Color Centers in Alkali Halides (Rome, 1968) Abstract 17.

[11] D. Y. Smith, Phys. Rev. **137A** (1965) 574.

[12] J. Margerie and R. Romestain, Comp. Rendus (Paris) **258** (1964) 4490.

[13] T. A. Fulton and D. B. Fitchen, Phys. Rev. **179** (1969) 846.

[14] L. Schiff, in: Quantum Mechanics (McGraw-Hill, New York, 1968). 3rd ed.

[15] G. Chiarotti, U. M. Grassano, G. Magaritondo and R. Rosei, Nuevo Cim. **B64** (1969) 159.

[16] S. E. Schnatterly, Phys. Rev. **140A** (1965) 1364.

[17] W. Gebhardt and K. Maier, Phys. Stat. Solidi **8** (1965) 303.

[18] R. K. Swank and F. C. Brown, Phys. Rev. **130** (1963) 34.

[19] M. Born and K. Huang, in: Dynamical Theory of Crystal Lattices (Oxford, 1954).

[20] H. Pick, Nuovo Cim. VII, (ser. X) **2** (1958) 498; H. Fedders, M. Hunger and F. Lüty, Phys. Chem. Solids, **22** (1961) 299.

[21] L. Bosi, P. Podini and G. Spinolo, Phys. Rev. **175** (1968) 1133.

[22] L. F. Stiles, M. P. Fontana and D. B. Fitchen, Solid St. Commun. **7** (1969) 681.

[23] L. F. Stiles, M. P. Fontana and D. B. Fitchen, Phys. Rev. **B2** (1970) 2077.

[24] U. Öpik and R. F. Wood, Phys. Rev. **179** (1969) 772; R. F. Wood, U. Öpik, Phys. Rev. **179** (1969) 783.

[25] H. S. Bennet, Phys. Rev. **16** (1968) 729; **184** (1969) 918; M. I. Petrashen, I. V. Abarenkov, A. A. Berezin and R. A. Evarestov, Phys. Stat. Solidi **40** (1970) 9, 433.

[26] B. S. Gourary and F. J. Adrian, Phys. Rev. **105** (1957) 1180; C. Laughlin, Solid St. Commun. **3** (1965) 55.

[27] T. Kojima, J. Phys. Soc. Japan **12** (1957) 918; R. F. Wood and H. W. Joy, Phys. Rev. **136A** (1964) 451.

[28] J. K. Kubler and R. J. Friauf, Phys. Rev. **140A** (1965) 1742.

[29] J. A. Krumhansl and N. Schwartz, Phys. Rev. **89** (1953) 1154; S. F. Wang, Phys. Rev. **132** (1963) 573.

[30] Refs. [2, 27]; see also E. Perlin, Usp. Fiz. Nauk. **6** (1964) 542.

[31] N. W. Mott and R. W. Gurney, in: Electronic Processes in Ionic Crystals (Oxford, 1940).

[32] H. Fröhlich, H. Pelzer and S. Zieman, Phil. Mag. **41** (1950) 221.

[33] J. H. Simpson, Proc. R. Soc. **A197** (1949) 269.

[34] W. B. Fowler, Phys. Rev. **135A** (1964) 1725.

[35] G. Iadonisi and B. Preziosi, Nuevo Cim. **B48** (1967) 92.

[36] S. F. Wang, Phys. Rev. **153** (1967) 939; R. L. Gilbert and J. J. Markham, J. Phys. Chem. Solids **30** (1969) 2699.

[37] L. D. Bogan, L. F. Stiles and D. B. Fitchen, Phys. Rev. Letters **23** (1969) 1495; L. D. Bogan and D. B. Fitchen, Phys. Rev. **B1** (1970) 4122.

[38] M. Bertolaccini, L. Bosi, S. Cova and C. Bussolati, in: Int. Symp. on Color Centers in Alkali Halides (Rome) **A68** Abstract 20.

[39] H. Kühnert and W. Gebhardt, (Ibid.) abstract 118.

[40] H. Ohkura, K. Imamaka, O. Kamada, Y. Mori and T. Iida, J. Phys. Soc. Japan **42** (1977) 1942.

[41] K. Imamaka, T. Iida and H. Ohkura, J. Phys. Soc. Japan **43** (1977) 519.

[42] R. E. Hetrick and W. D. Comptom, Phys. Rev. **155** (1967) 649.

[43] G. Spinolo and F. C. Brown, Phys. Rev. **135A** (1964) 450.

[44] F. Lüty, in: Physics of Color Centers, ed. W. B. Fowler (Academic Press, New York, 1968).

[45] L. Bosi, S. Cova and G. Spinolo, Phys. Stat. Solidi (b)**68** (1975) 603.

[46] H. Kühnert, Phys. Stat. Solidi **21** (1967) K171; H. Kühnert, Thesis, Frankfurt (1978) (unpublished).

[47] M. P. Fontana and D. B. Fitchen, Phys. Rev. Letters **23** (1969) 1497.

[48] M. P. Fontana, Phys. Rev. **B2** (1970) 4304.

[49] G. Baldachini, U. M. Grassano and A. Tanga, J. Phys. C: Solid St. Phys. **7** (1976) 154; Sol. St. Commun. **21** (1977) 225.

[50] Y. Kondo and H. Kanzaki, Phys. Rev. Letters **34** (1975) 664.

[51] K. Park and W. L. Faust, Phys. Rev. Letters **17** (1966) 137.

[52] S. Wang, M. Matsuura, C. C. Wong and M. Inoue, Phys. Rev. **B7** (1973) 1695.

[53] F. S. Ham, Phys. Rev. **B8** (1973) 2926.

[54] Y. Kayanuma and Y. Toyozawa, J. Phys. Soc. Japan **40** (1976) 355.
[55] F. S. Ham and U. Grevsmuhl, Phys. Rev. **B8** (1973) 2945.
[56] Y. Kayanuma, J. Phys. Soc. Japan **40** (1976) 363.
[57] L. F. Mollenauer and G. Baldachini, Phys. Rev. Letters **29** (1972) 465.
[58] I. Schneider, in: Int. Conf. on Defects in Insulating Crystals (Gatlinburg, USA, 1977).
[59] T. Fulton and D. B. Fitchen, Phys. Rev. **B1** (1970) 4011.
[60] Y. Merle d'Aubigne and A. Roussel, Phys. Rev. **B3** (1971) 1421; P. Edel, G. Henies, Y. Merle d'Aubigne, R. Romestain and Y. Twarowski, Phys. Rev. Letters **28** (1972) 1268; P. Edel, Y. Merle d'Aubigne and R. Louat, J. Phys. Chem. Solids **35** (1974) 67.
[61] C. B. Harris, M. Glasbeck and E. B. Hussley, Phys. Rev. Letters **33** (1974) 537.
[62] N. V. Karlov, J. Margerie, Y. Merle d'Aubigne, J. Physique **24** (1963) 717.
[63] H. Panepucci and L. F. Mollenauer, Phys. Rev. **178** (1968) 589.
[64] M. P. Fontana, Phys. Rev. **B2** (1970) 1107; C. H. Henry, Phys. Rev. **140A** (1965) 256.
[65] E. S. Sabisky and C. H. Anderson, Phys. Rev. **B1** (1970) 2028.
[66] H. Seidel and E. Wolf, in: Physics of Color Centers (op. cit).
[67] R. W. Warren, D. W. Feldman and J. G. Castle, Phys. Rev. **136A** (1964) 1357.
[68] J. H. Van Vleck, Phys. Rev. **57** (1940) 426.
[69] R. De L. Kronig, Physica **6** (1939) 33.
[70] B. R. McAvoy, D. W. Feldman, J. G. Castle Jr. and R. W. Warren, Phys. Rev. Letters **6** (1961) 618.
[71] D. Schmid and V. Zimmerman, Phys. Letters **27A** (1968) 459.
[72] L. F. Mollenauer, S. Pan and S. Yugverson, Phys. Rev. Letters **23** (1969) 683.
[73] L. Porret and F. Lüty, Phys. Rev. Letters **26** (1971) 843.
[74] L. F. Mollenauer, S. Pan and A. Winnacker, Phys. Rev. Letters **26** (1971) 1643.
[75] P. D. Parry, T. R. Carver, S. O. Sari and S. E. Schnatterly, Phys. Rev. Letters **22** (1969) 326.
[76] Y. Ruedin and F. Porret, Helv. Phys. Acta **41** (1968) 1294.
[77] Y. Ruedin, P. A. Schwegg, C. Jaccard and M. Aegerter, Phys. Stat. Solidi (b)**54** (1972) 565; **55** (1973) 218.
[78] P. A. Schwegg, C. Jaccard and M. Aegerter, Phys. Letters **A42** (1973) 369.
[79] A. Mielich, Z. Phys. **176** (1963) 168.
[80] F. Lüty, Haebleiterprobleme VI, (F. Wieneg, Brauschweig, 1961).
[81] D. Schmid and H. C. Wolf, Z. Phys. **170** (1972) 455.
[82] J. A. Strozier and B. G. Dick, Phys. Stat. Solidi **31** (1969) 203.
[83] R. S. Knox and K. Teegarden, in: Physics of Color Centers 1968 (op. cit.); M. N. Kabler, in: Radiation Damage Processes in Materials, ed. C. H. S. Dupuy (Noordhoff-Leyden, 1975).
[84] I. L. Kuusmann, P. Kh. Liblik and Ch. B. Luschik, Sov. Phys. JETP Letters **21** (1975) 72; I. L. Kuusmann, P. Kh. Liblik, G. G. Liid'ya, N. E. Luschik, C. B. Luschik and T. A. Soovik, Sov. Phys. Solid St. **17** (1976) 2312; Ch. B. Luschick, I. Kuusmann, G. Liid'ya, N. E. Luschick, V. G. Plakhanov, A. Ratas, T. Soovik and P. Liblik, J. Luminesc. **11** (1976) 285.
[85] T. Hayashi, T. Ohata and S. Koshino, J. Phys. Soc. Japan **42** (1977) 1647.
[86] M. N. Kabler and D. A. Patterson, Phys. Rev. **136A** (1964) 1296; R. B. Murray and P. J. Keller, Phys. Rev. **137A** (1965) 942.
[87] D. Pooley and W. A. Runciman, J. Phys. C: Solid St. Phys. 3 (1970) 1815.
[88] R. G. Fuller, R. T. Williams and M. N. Kabler, Phys. Rev. Letters **25** (1970) 446; R. T. Williams and M. V. Kabler, Phys. Rev. **B9** (1974) 1897.
[89] A. E. Purdy, R. B. Murray, K. S. Song and A. M. Stoneham, Phys. Rev. **B15** (1977) 2170.
[90] J. U. Fischbach, D. Frölich and M. N. Kabler, J. Luminesc. 6 (1973) 29.
[91] T. Karasawa and M. Hirai, J. Phys. Soc. Japan **40** (1976) 128.
[92] M. J. Marrone and M. N. Kabler, Phys. Rev. Letters **27** (1971) 1283.
[93] M. J. Marrone and M. N. Kabler, Phys. Rev. Letters **31** (1973) 467.
[94] A. Wasiela, G. Ascarelli and Y. Merle d'Aubigne, Phys. Rev. Letters **31** (1973) 993.
[95] W. B. Fowler, M. J. Marrone and M. N. Kabler, Phys. Rev. **B8** (1973) 5909.

[96] J. A. Ramamurti and K. J. Teegarden, Phys. Rev. **145** (1966) 698.
[97] A. Hattori, M. Tomura, O. Fujii and H. Nishimura, J. Phys. Soc. Japan **41** (1976) 194.
[98] T. Hayashi, T. Ohata and S. Koshina, J. Phys. Soc. Japan **42** (1977) 1647.
[99] A. Sumi and Y. Toyozawa, J. Phys. Soc. Japan **35** (1973) 137.
[100] G. Guillot, E. Mercier and A. Nouilhat, J. Phys. Letters **38** (1977) L495.
[101] Y. Suzuki and M. Hirai, J. Phys. Soc. Japan **43** (1977) 1679.
[102] H. R. Phillip and H. E. Ehenreich, Phys. Rev. **131** (1963) 2016; G. Baldini and B. Bosacchi, Phys. Rev. **166** (1968) 863.
[103] A. Onton, P. Fisher, and A. K. Ramdas, Phys. Rev. **163** (1967) 868; B. B. Kosicki and W. Paul, Phys. Rev. Letters. **17** (1966) 246.
[104] H. Kaplan, J. Phys. Chem. Solids **24** (1963); 1593; L. Liu and D. Brust, Phys. Rev. **157** (1967) 627.
[105] J. M. Blatt and F. Weisskopf, in: Theoretical Nuclear Physics (Wiley, New York, 1958).
[106] F. Bassani, G. Iadonisi and B. Preziosi, Phys. Rev. **186** (1969) 735.
[107] Y. Onodera and Y. Toyozawa, J. Phys. Soc. Japan **22** (1967) 833; A. B. Kunz, Phys. Rev. **B4** (1971) 609.
[108] F. Lüty, F. Phys. **160** (1960) 1.
[109] N. Inchauspe, Phys. Rev. **106** (1957) 898; F. Nakazawa and H. Kanzaki, J. Phys. Soc. Japan **22** (1967) 844; M. Hirai and M. Ikezawa, J. Phys. Soc. Japan **22** (1967) 810.
[110] G. Chiarotti and U. M. Grassano, Nuevo Cim. **46B** (1966) 78.
[111] S. Benci and M. Manfredi, Solid State Commun. **9** (1971) 1255.
[112] R. Hilsch, Z. F. Phys. **44** (1927) 860; Proc. Phys. Soc. (London) **49** (1937) 40.
[113] F. Seitz, J. Chem. Phys. **6** (1938) 150.
[114] D. Bramanti, M. Mancini and A. Ranfagni, Phys. Rev. **B3** (1971) 3670.
[115] P. H. Yuster and C. J. Delbecq, J. Chem. Phys. **21** (1953) 892.
[116] M. Forro, Z. Phys. **56** (1929) 534.
[117] U. Giorgiani, G. Montio, G. Saitta and G. Vermiglio, Phys. Letters **54A** (1975) 45; U. Giorgiani, G. Mondio, P. Perillo, G. Saitta and G. Vermiglio, Phys. Rev. **B15** (1977) 5983.
[118] G. W. King and J. H. Van Vleck, Phys. Rev. **56** (1939) 464.
[119] E. U. Condon and G. H. Shortley, in: The theory of Atomic Spectra (Cambridge,
[120] S. Sugano, J. Chem. Phys. **36** (1962) 122.
[121] F. E. Williams, B. Segall. P. D. Johnson, Phys. Rev. **108** (1957) 46.
[122] Y. Toyozawa and M. Inoue, J. Phys. Soc. Japan **21** (1966) 1663.
[123] A. Honma, J. Phys. Soc. Japan **24** (1968) 1082.
[124] K. Cho, J. Phys. Soc. Japan **27** (1969) 646.
[125] V. Grasso, P. Perillo, G. Vermiglio, Solid St. Commun. **11** (1972) 563.
[126] D. Lemoyne, J. Duran, M. Billardon and Le Si Dang, Phys. Rev. **B14** (1976) 747.
[127] C. C. Klick and J. H. Schulman, Solid St. Phys. **5** (1957) 97; R. Edgerton and K. Teegarden, Phys. Rev. **136A** (1964) 1091; C. C. Klick and W. D. Compton, J. Phys. Chem. Solids **7** (1958) 170; A. Fukuda, S. Makishima, T. Mabucchi and R. Onara, J. Phys. Chem. Solids **28** (1967) 1763.
[128] A. Fukuda, S. Makishima, T. Mabuchi and R. Onaka, J. Phys. Chem. Solids **28** (1967) 1763.
[129] M. F. Trinkler and I. S. Zoloukina, Izv. Acad. Nauk. SSR. **40** (1976) 1939; Phys. Stat. Solidi **B79** (1977) 49.
[130] S. Benci, M. P. Fontana and M. Manfredi, Solid St. Commun. **18** (1976) 1423.
[131] A. Fukuda, Phys. Rev. **B1** (1970) 4161.
[132] A. Ranfagni, Phys. Rev. Letters **28** (1972) 743.
[133] A. Ranfagni and G. Viliani, Phys. Rev. **B9** (1974) 4448.
[134] M. P. Fontana, G. Viliani, M. Bacci and A. Ranfagni, Solid St. Commun. **18** (1976) 1615.
[135] M. P. Fontana and J. A. Davis; Phys. Rev. Letters. **23** (1969) 974.
[136] A. Fukuda, K. Cho, and H. J. Pauss, in: Luminescence of Crystals, Molecules and Solutions, ed. F. Williams (Plenum, New York, 1973) p. 478.

[137] M. Bacci, B. D. Bhattachakyya, A. Ranfani and G. Viliani, Phys. Letters. **55A** (1976) 489.
[138] J. M. Donahue and K. Teegarden, J. Phys. Chem. Solids **29** (1968) 2141.
[139] Le. Si. Dang, R, Romestain, Y. Merle d'Aubigne and A. Fukuda, Phys. Rev. Letters. **38** (1977) 1539.
[140] W. D. Drotning and H. G. Drickamer, Phys. Rev. **B13** (1976) 4568.
[141] M. Bacci, A. Ranfagni, M. P. Fontana, and G. Viliani, Phys. Rev. **B11** (1975) 3052.
[142] U. Giorgiani, V. Grasso and P. Perillo, Phys. Rev. Letters **23** (1969) 640; U. Giorgiani, V. Grasso and G. Saitta, Nuevo Cim. **B68** (1970) 100.
[143] A. Ranfagni, G. P. Passi, P. Farberi, M. Bacci, M. P. Fontana and G. Villiani, Phys. Rev. Letters. **35** (1975) 752.
[144] A. Ranfagni, Phys. Letters **62A** (1977) 395.
[145] L. D. Landau and E. M. Lifshitz, in: Mecanique (MIR, Moscow., 1966) Ch. 7.
[146] A. Ranfagni and G. Viliani, Sol. St. Commun. (in Press).
[147] U. Grasso, P. Perillo and G. Vermiglio, Nuovo Cimento **13B** (1973) 42.
[148] R. Illingworth, Phys. Rev. **136A** (1964) 508.
[149] A. Ranfagni, G. Viliani, M. Letica, G. Malesini, Phys, Rev. **B16** (1977) 890.
[150] K. Huang and A. Rhys, Proc. R. Soc. **A204** (1950) 406.
[151] J. M. Ortega, Ph.D., Université Paris Sud. (1977).
[152] W. D. Compton and C. C. Klick, Phys. Rev. **112** (1958) 1620.
[153] H. Seidel, Phys. Letters **7** (1963) 27.
[154] I. Schneider and M. E. Caspari, Phys. Rev. **133** (1964); 1193; M. Ikezawa, J. Phys. Soc. Japan **19** (1964) 529; R. T. Mc Call and L. T. Grossweiher, J. Appl. Phys. **38** (1967) 284; H. Pick and D. Haarer, Z. Phys. **200** (1967) 213.
[155] H. Engström, Phys. Rev. **B11** (1975) 1657.
[156] Y. Farge, Cze. J. Phys. **B20** (1970) 611.
[157] J. M. Ortega, Phys. Rev. **B16** (1977) 3782.
[158] J. M. Ortega, P. Lagarde and Y. Farge, Solid St. Commun. **16** (1975) 957.
[159] J. M. Ortega, Commun. Phys. **2** (1977) 45.
[160] S. P. Mc Glynn, T. Azumi and M. Kinoshita, in: Molecular Spectroscopy of the Triplet State (Prentice Hall, New Jersey., USA, 1969).
[161] Y. Farge, J. M. Ortega and R. H. Silsbee, J. Chem. Phys. **69** (1978) 3972.
[162] J. Jortner, S. A. Rice and R. M. Hochstrasser, Adv. Photoch. **7** (1969) 149.
[163] M. Bixon and J. Jortner, J. Chem. Phys. **48** (1968) 715.

5 | APPLICATIONS

The knowledge of color centers discussed in the preceding chapters can be used to study more specific problems and to present some applications. In this chapter we shall first discuss the creation of color centers by ionizing radiation in ionic crystals; our understanding of this very general phenomenon has in fact improved considerably in the last five years and it is now well established that the primary process of defect formation is the formation of an F- and H-center pair via an excitonic recombination. Recent experiments have given detailed information on the fundamental mechanism which is still not completely understood; secondary processes related to interstitial or vacancy mobility are however well understood now and will be discussed here in some detail.

Color centers can also be used for some applications. For example, defect creation by irradiation can be used for information storage by irradiating with an electron beam; for radiation dosimetry, by measuring the F-center concentration or the thermoluminescence resulting from their thermal bleaching; such dosimetry may be used for instance to yield information about the temperature of geological samples at the time the radiation damage had occurred. We have seen also that it is possible to change the optical absorption spectra of a sample containing color centers by the appropriate optical bleaching, or again to induce dichroism by bleaching with polarized light; also in these cases there are possibilities for high capacity information storage. Para-electric defects may be used in several instances, for instance in the measurement of very low temperatures by the Kerr effect. Finally, color centers can be used to make tunable lasers over a large wavelength range, especially on the near infrared ($1\mu \rightarrow 3\mu$) where such lasers are missing.

5.1. Creation of color centers by ionizing radiations [1–5]

All ionizing radiations can create F-centers in alkali-halides: γ-rays, X-rays, electrons; also UV phonons if their energy is greater than the first excitonic transition in the crystal. It is now well established that the primary defects produced by these irradiations are Frenkel pairs of F- and H-centers. Phenomena related to the creation of defects are complex because the efficiency of F-center formation is strongly dependent on temperature and impurity content; in particular some impurities decrease the formation rate while others increase it. Much care and patience were necessary to separate all the observed effects and now the formation of color centers in alkali-halides by ionizing radiations is quite well understood. Thus it is possible to identify two main temperature ranges: the low temperature range in which the H-center formed is stable (T < about 50 K), and the high temperature range in which the interstitial is mobile. For simplicity we shall call the primary process that which takes place in the low temperature range, whereas the secondary processes take place in the higher temperature range.

5.1.1. The primary effects

It is now established that the formation of Frenkel pairs in alkali-halides is due to excitonic recombination. The clearest evidence is given by the following experiment [6]: in a crystal KCl:Pb X-ray irradiated at 77 K self-trapped holes are formed (V_k-centers) and also Pb^+ ions due to electron capture by the Pb^{2+} ions. It is possible to free these electrons by optical irradiation; if this is done at 4 K, the formation of F- and H-centers by electron self-trapped hole recombination is observed.

The formation of free interstitials.   To study the excitonic recombination at low temperatures it is possible to follow either the F- or the H-center creation. Let us first discuss an experiment which yields the efficiency of formation of isolated H-centers [7–9]. A crystal is previously colored at room temperature to yield F_2-centers, and then it is irradiated at lower temperature; one observes a decrease in F_2-center concentration caused by the trapping of the H-centers mobile at the temperature of the second irradiation. From this decrease, it is possible to measure the efficiency $\dot{\varepsilon}$ of the free H-center formation which is found to vary, as shown in fig. 5.1, according to the equation:

$$\dot{\varepsilon} \propto \exp(-W/kT). \tag{5.1}$$

The same result is obtained if interstitial trapping is observed on another

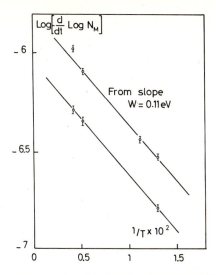

Fig. 5.1. The primary process creation yield for F-centers versus $1/T$ in LiF (log–log scale). In two previously colored crystals the yield is measured by the bleaching rate of the F_2-centers which trap the interstitials created during irradiation (after Durand et al. [7]).

defect. We thus reach the conclusion that the efficiency of the free H-center formation is itself thermally activated. This activation energy can come either from the primary process, i.e. from the chemical reaction during the excitonic recombination or from the decorrelation of F- and H-centers which, interacting each other at least elastically, may need an activation energy.

Ioth and Saidoh [10] have proposed that this activated process comes from the decorrelation of the two defects because it is well known that at temperature, where H-centers are not mobile, correlated F–H centers are easily observed; having observed that activation energies of table 5.1, are lower than the activation energies of free H-centers, they came to the conclusion that the mobile entity is the H-center in an electronic excited state and not in its groundstate and that the primary photochemical reaction creates

Table 5.1
The activation energy of formation of free H-centers [7, 8, 9].

LiF	0.11 eV
NaCl	0.07 eV
RbCl	0.06 eV
KCl	0.07 eV
KBr	0.03 eV
KI	0.01 eV

excited H-centers and F-centers in their groundstate. These conclusions are strongly supported by pulsed experiments with electrons using a Febetron: in that case, one can observe, after a pulse of a few nanoseconds an immediate decrease of F- and H-centers in some microseconds coming from their combination [11].

Hall et al. [6] have recently performed an experiment which is in agreement with the Itoh's idea of a migration of the H-center in its excited state. In KCl:Pb, they have formed F- and H-centers by detrapping of electrons as described previously and trapping on optically aligned V_k-centers. They have not observed any dichroism in the H-absorption band which indicates that after its migration in a [110] direction, the H-center in its excited state can rotate easily before relaxing to its groundstate where the rotation is not so easy.

Before discussing the excitonic mechanism, it is interesting to make some remarks about the geometry of H-center diffusion. Because of the nature of the V_k-center self-trapped in a $\langle 110 \rangle$ direction it is expected that the H-center formed by $e-V_k$ recombination is diffusing along the direction of the V_k-center (replacement collision chains). Let us just present an interesting experiment which is in a quite good agreement with this assumption [12]. In KCl, KBr and NaCl crystals, Schmid et al. have investigated at 77 K excitonic recombination by four-photon absorption with a Q-switched ruby laser and they have measured the species ejected from the crystal in the vacuum. As expected halogen ions and a much larger number of halogen atoms are observed, but they are not ejected isotropically: a directional emission is observed in the $\langle 110 \rangle$ and $\langle 211 \rangle$ directions during interaction of the crystal with strong laser pulses. This result is in good agreement with replacement collision chains after a non-radiative decay of the exciton. The $\langle 211 \rangle$ directions were quite unexpected and can be explained in terms of defocussing of the $\langle 110 \rangle$ momentum because $\langle 211 \rangle$ directions in alkalihalides are also halogen rows as well as $\langle 110 \rangle$ [13]. A similar sputtering has been observed by Townsend et al. [14] in NaCl irradiated with low energy electrons at low temperature instead of room temperature as in the previous case. Some differences have been observed which probably result from the irradiation temperature as well as differences due to the fact that only the first exciton state is populated with the laser pulse when electrons can also produce higher excitons states. Itoh [5] has discussed what can be learnt from these experiments on the dynamic motion of interstitials.

The non-radiative decay of the exciton. We have discussed in some detail in chapter 4, free and self-trapped excitons. We have said that a large fraction of the exciton energy is released by non-radiative transitions and amongst them, one gives rise to the Frenkel pair formation. The efficiency of non-radiative transitions is so high that they certainly take place in the

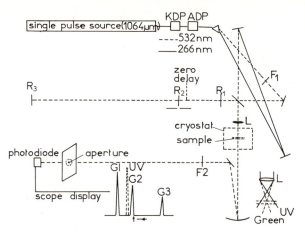

Fig. 5.2. Apparatus for measurement on a picosecond timescale of absorption at 532 nm following excitation at 266 nm (after Bradford et al. [15]).

picosecond range and it is necessary to study the F-center formation in this time scale. This experiment has been performed by Bradford et al. [15] with an apparatus described in fig. 5.2.

In this experiment, a mode-locked Nd:YAG laser generates picosecond pulses at 1.064 μm. The second harmonic (532 nm) and the fourth harmonic (266 nm) are generated by non-linear crystals. The UV pulse is sent to the crystal along a fixed delay path and creates electron–hole pairs in the sample by two-photon absorption. The green pulse, which is used to monitor the F-center absorption in KCl, is attenuated by the filter F_1 before being directed down a variable optical delay line comprising partial reflectors R_1, R_2 and R_3. The three green pulses are then sent to the crystal collinearly with the UV pulse and the F_2 filter transmits only the green light. The green interrogation pulses are recorded with fast vacuum photodiodes and oscilloscopes. The three pulses are used to monitor the absorption of the crystal before the UV pulse, after a given constant time (pulse G3) and after another given time (pulse G2). The results obtained by Bradford et al. are shown in fig. 5.3. They observed that F-centers are produced exponentially with time with a typical time of 11 ps (solid line of fig. 5.3). This very interesting experiment shows that the F-center is created in its groundstate and confirms, for this point, the experiments with electron pulses [11]. Their results also make possible the determination of the efficiency production of a Frenkel pair during an excitonic recombination which is found to be about 15%. Because of the very short time of the F-center formation, these authors assume that the F-center formation takes place from fairly high-lying electronic levels of the self-trapped exciton. This result is confirmed by a recent experiment performed by Williams [16]; in this experiment STE metastable states

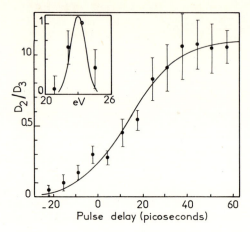

Fig. 5.3. Change of a KCl crystal absorption at 532 nm versus delay after the 266 nm pulse. The full line is a best-fit convolution of pulse shapes with an exponential form for defect production, yielding $\tau \approx 11$ psec. (after Bradford et al. [15]).

are formed by electrons pulses and F-center formation is observed when the STE is excited from its ground triplet state to the second excited triplet state $^3\Sigma_g$ (see fig. 4.18).

The question which remains open now is the process of the formation of a Frenkel pair with the F-center in its groundstate from a singlet state of the self-trapped exciton. Let us just explain the model proposed by Toyozawa [17] which is in competition with other models [4, 10] which are also very reasonable given the actual amount of experimental information. In the following, we reproduce a large part of his 1974 paper. A pulsed irradiation produces holes and electrons. Holes are immediately self-trapped (V_k-centers) while electrons lose kinetic energy by different processes and then are trapped by the V_k-centers. The electrons initially trapped at shallow levels will cascade down by emitting phonons within fairly short times. We said previously that the electron can be seen as fairly delocalized around the V_k-center and then can be described by hydrogenoid states. When the electron has reached the $2p_z$ state (z along the V_k-axis) it takes time to reach the $1s$ state which is about 2 eV below. Then Toyozawa assumes that this $2p_z$ singlet state is the initial state of the photochemical reaction. Now to describe the coupling of this $2p_z$ state with the two ions in the V_k-center, Toyozawa uses another but equivalent wavefunction. In the V_k-center, the two central halide ions are situated much closer to each other than in a regular lattice, forming an X_2^- molecule; then the extra electron is expecting the attractive potential of a pair of halved anion vacancies (each with effective charge $+e/2$) like an F_2^+-center. Then, instead of the atomic

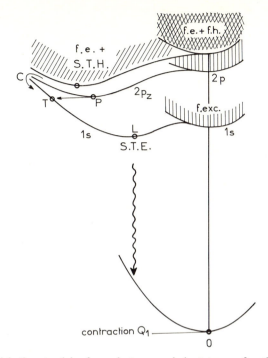

Fig. 5.4. The adiabatic potentials of one-electron one-hole states as a function of contracting mode Q_1 of a nearest pair of anions (after Toyozawa [17]).

picture, he uses a molecular picture and describes the $2p_z$ state as a linear combination of one-electron vacancy orbitals. But this state is expected to be unstable: if the X_2^- molecule moves towards the $+z$ direction (fig. 5.4) a relative increase of the amplitude of the vacancy orbital appears on the opposite side. When X_2^- gets to and occupies the normal anion site which was initially a halved vacancy, the electron will be entirely localized on the opposite side where it now finds a complete vacancy and where it will be in a 1s state because the $2p_z$ state can be described with a linear combination of 1s vacancy orbitals. Now we have an adjacent pair of F^- (in its ground-state) and H-centers. However, the effective charge of the vacancy is increased from $+1/2$ to 1 and so the binding energy by a factor of 4. If this increase in electronic binding energy is sufficiently greater than the increase in lattice energy, the adiabatic potential is expected to have steep descents toward both directions of X_2^- away from its symmetry position as shown by the thick full line of fig. 5.4. Then, the H-center has the kinetic energy to travel in the direction of the V_k-center. As shown in fig. 5.4, during the relaxation of the $2p_z$ state a non-radiative transition can occur to the triplet groundstate and the closed singlet state. Then the relaxation of the $2p_z$ state has two competing channels, one leading to color center formation, the other to intrinsic luminescence as is observed.

This model is certainly crude[†] and we can expect important developments in the near future. We discussed it in some length because we feel that the dynamics of excited states is a very important subject. A lot of photo-chemical reactions are known in alkali-halides and we do not know how they take place. For example, an optical excitation of a U-center brings the neutral halogen atom in an interstitial position [18], an optical excitation of substitutional OH^- brings the formation of interstitial $H°$ and $O°$ and an F-center [19]. It has also been shown in KCl that the migration energy of F-centers in their excited states is 0.13 eV instead of the 1.5 eV in their groundstate [20].

5.1.2. The secondary effects

As the temperature increases a great number of secondary effects may take place. For instance, in KCl the H-centers become mobile at about 50 K and the V_k-centers at about 150 K; these effects may alter the efficiency of the primary effect. Starting from 240 K, the F^+-centers become mobile and may recombine with electron centers or with trapped interstitials. Finally, the F-centers themselves become mobile at about 500 K. Among these several effects, only the interstitial–F-center recombination (which restores the perfect crystal) and the recombination of F^+ with either F- or F_2-centers are effectively understood.

The reader will perhaps remember that the H_A-center created at 77 K is an H-center stabilized by a monovalent positive impurity. This defect is already an example of defects related to the H-center diffusion. At higher temperatures, these atoms form agglomerates the size of which increases with irradiation temperature. Upon irradiation the interstitials will have a finite probability of recombining with F-centers before being stabilized on one such cluster. The creation rate for F-centers will thus decrease contin-uously with irradiation time. In very pure samples, most authors have observed that the number of F-centers increases as $\Phi^{1/2}$ near room tempera-ture, where Φ is the incident energy flux. The preceding model yields a correct interpretation of this law [7, 22]. Furthermore, the number of free

---

[†] For example, the small dip at $2p_z$ configuration is an *ad hoc* drawing to explain the very strange effect observed by Karazawa and Hirai [21] which is a decrease by a factor five of the efficiency of the F-center formation when the sample temperature is lowered from 4.2 to 1.6 K.

interstitials under irradiation may be measured: in fact, the F_2-centers are formed only above 240 K in most alkali-halides; in a crystal irradiated at 300 K, containing F_2-centers and further irradiated below 240 K, one observes the destruction of these centers by trapping of free interstitials. The destruction rate for these centers is proportional to their concentration. This method allows the direct measurement of the primary process efficiency as a function of temperature or nature of the irradiation (fig. 5.1). Sonder [23] has shown that this efficiency is very high since in KCl at 200 K one phonon of 1 MeV (^{60}Co) created about 2.4×10^4 Frenkel pairs, i.e. it takes about 40 eV to form such a pair with incident radiation.

Upon irradiation, F-centers may be ionized and form F^+-centers which will be mobile at room temperature. Upon association with an F-center, an F_2^+ will be formed, and possibly, after electron capture, an F_2-center: the reaction kinetics will be: [24, 25]

$$F + F^+ \rightarrow F_2^+ \quad \text{(a)} \tag{5.2}$$
$$F_2^+ + e^- \rightarrow F_2 \quad \text{(b)}$$

with similar reactions the F_3-centers are formed:

$$F_2 + F^+ \rightarrow F_3^+ \quad \text{(a)} \tag{5.3}$$
$$F_3^+ + e^- \rightarrow F_3 \quad \text{(b)}.$$

If the irradiation takes place at sufficiently high temperatures, very large clusters of F-centers may be formed. Thus, in LiF irradiated with X-rays at 600 K we have observed by X-ray scattering clusters with mean dimensions of 14.6 Å, i.e. containing approximatively 820 F-centers each. The formation process of complex electron centers (F_n) is now essentially understood. However, some questions still await an answer: why is the activation energy for the reaction [5.2(a)] much smaller than the migration energy of the F^+-center? Why is the activation energy of reaction [5.3(a)] higher than that for reaction [5.2(a)]? Schneider [26] indirectly has measured the activation energy of the F^+-center in KCl and he finds exactly the energy measured for the reaction (5.2), i.e. less than 40% of the energy obtained at high temperatures from classical diffusion measurements (diffusion velocity of K in KCl). This very surprising result would indicate that the concept of activation energy would be valid only in a limited temperature domain, and that such an energy may be a function of temperature. It would be useful to investigate in more detail the vacancy (F^+) jump mechanism in order to understand such an effect.

The phenomena connected with the motion of the V_k-centers are less well established. Pooley [27] has suggested that a hole–electron recombination involving a hole trapped at an impurity site might not yield a Frenkel pair, i.e. that the creation rate of F-centers must diminish as the V_k-centers become mobile. In doped KI and KCl Dawson and Pooley [28] have effectively observed a decrease of the creation rate at the temperature at which the V_k-center becomes mobile. In pure crystals [7, 23] this effect is not observed and the creation rate goes on increasing with temperature.

The impurity related effects are numerous. Although it seems established that they little modify the primary creation rate, they may increase the creation rate at room temperature by interstitial trapping, and may decrease it by trapping holes. Molecular impurities may equally be dissociated to yield F-centers [19].

We shall not discuss radiolysis processes in other materials here which have not been as well studied as the alkali-halides. In conclusion, it is important to notice that the knock-on process can also produce defects in ionic crystals. This process is well known in alkaline earth oxides [29] and has been observed in LiF through the correlated formation of Li interstitials upon irradiation with neutrons or electrons [30].

5.2. The use of color centers in X- and γ-ray dosimetry [31, 32]

Most ionic crystals irradiated at room temperature with ionizing radiations emit light as they are heated after the irradiation. This thermoluminescence phenomenon is well known and the related techniques are simple: the crystal temperature is increased at constant heating rate v, and the emitted light $I_v(T)$ detected with a photomultiplier as a function of temperature. Adding interference filters or better a monochromator, the spectral distribution $I(v)$ of the thermoluminescence may be determined. The cause of this phenomenon is well understood: the irradiation creates charges which are finally trapped on a given site; above a certain temperature these charges are thermally ejected, become mobile and recombine at various defects to emit luminescence. In general, thermoluminescence involves a large number of parameters, making the interpretation of the effect difficult; in particular the influential parameters are the different type of impurities and their concentration in the crystalline lattice. However it is evident that the emission intensity is directly related to the irradiation dose; this yields dosimeters of high sensitivity, if we consider the high detectivity of photomultipliers [31]. Thus, for instance, CaF_2:Mg has excellent sensitivity and a linear response between 10^{-3} R and 10^5 R [33]. The samples are generally reduced to powder form and package in a transparent substance to insure easy manipulation. These dosimeters are re-usable since the defects created may be bleached by heating at a sufficiently high temperature; there is however

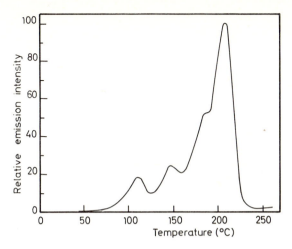

Fig. 5.5. Thermoluminescence spectrum of LiF:Mg (after Klick et al. [34]).

an aging effect, still not well understood, which limits the number of possible cycles.

In order to produce personal dosimeters, it is advisable to choose a substance with an electron density approximatively equivalent to that of the human body, in order to obtain information on the radiations effects which are as close as possible to the situation in the human tissues. Thus a considerable effort has been expended to use LiF as a thermoluminescent dosimeter. This type of crystal, doped with Mg, is now widely employed even though the thermoluminescence mechanism is still not well known. These phenomena are in effect very complicated: in a γ-irradiated crystal, five thermoluminescence peaks are observed (fig. 5.5) in this figure the first lowest temperature peak is not shown). The highest temperature peak is used for dosimetry. Klick et al. [34] have shown that the spectral distribution of the luminescence associated to the five peaks is due to the F-center–V_k-center recombination; they have in fact used the blue thermoluminescence appearing at $-150°C$ in crystals irradiated at liquid nitrogen temperature. At $-196°C$, the irradiation creates F- and V_k-centers which become mobile at $-150°C$; at this temperature they will recombine with the F-centers and thus yield the thermoluminescence. If in LiF:Mg we know the nature of the moving charges and of the light emitting center, we still have no tight models for the defect which traps the hole [35, 37].

Fast neutrons may also be detected using F-centers in MgO [38]. In this detector, F-centers are formed by the knock-on process; the detection of the small amount of F-centers formed in association with hydrogen is accomplished by measuring the electrons escaping from these defects and from the crystal when it is heated at 350°C. This detector is energy independent from 2 to 15 MeV.

Thermoluminescence has been equally used to monitor rapidly and conveniently UV radiation on the earth's surface (λ > 320 nm). McCullough et al. [39] have used natural fluoride for this purpose. Their dosimeter is particularly designed to measure energy doses in the spectral region responsible for tanning of the skin.

5.3. Color centers in natural crystals: geological applications [40†, 41]

In natural minerals, a natural irradiation may exist due to the presence of radioactive traces which will produce defects by changing impurity charges states or producing F-centers and their agglomerates. In general, such defects are complex and are made up by a color center associated with some impurity. Such defects have been generally considered as curiosities leading to such effects as thermoluminescence and thermal and optical bleaching [42]. Alkali-halides are not very abundant in nature, but their coloration may in general be attributed to the defects discovered in their synthetic counterparts in the laboratory. In the particular case of halite (NaCl), the most common coloration is blue and is caused by the colloids formed by F-center aggregation; the minerals in the veins are generally at a temperature between room temperature and 100°C, and this favors the agglomeration of F-centers. Fluorite (CaF_2) is an abundant mineral which comes in many colors: red, green, yellow, blue, mauve or even colorless. Impurities play an important role in this case, making the interpretation of the optical absorption spectra very complicated. Aside from the green coloration due to the Sm substitutional impurity [43], the other colors are connected with complex centers whose structure is not known. Silica, equally abundant, is also sensitive to natural radiation and we also find amethyst and smoky quartz (in this case the defect is associated with Al which replaces Si). Color centers may appear in more complex minerals such as the calcites, the silicates etc. In most cases, the irradiation must have taken place during crystallization, since the more active rays are the weakly penetrating α-rays [44]. The crystallization takes place for most minerals at moderate temperature (e.g. 100 to 200°C for fluorite) in "hydrothermal solutions" and these samples often appear zoned; this is due to the variations in the radioactivity of water.

Color centers created by irradiation in natural ionic solids may be used

† Review paper.

to date the material absolutely or else they form a geological thermometer. For the first application [45] one measures the optical absorption related to color centers formed continuously in a material which contains long lived radioactive substances. Knowing the concentration of these substances and their efficiency for creation of color centers, it is then possible to date the mineral. A good match has in fact been observed between the age of a mineral vein and the coloration of the fluorites contained therein. This method must be used with caution since a possible slight heating may have partially bleached the color centers in the sample, thus falsifying the measured age. When the mineral contains color centers without containing radioactive impurities, it may be used as a geological thermometer. By measuring the thermal bleaching kinetics of yellow fluorites, Calas et al. [46] have observed first-order kinetics corresponding to the dissociation of the defect responsible for the color. By extrapolating these results, Calas has even determined the maximum temperature reached by the mineral vein, this temperature becoming smaller with the vein age. Thus this system makes up a geological thermometer. In the case studied by Calas, the vein was approximatively 500 million years old and its temperature never exceeded 110° C. This technique may also yield the formation time of the vein: in fact, fluorites as other minerals contain two-phase inclusions (liquid–gas), made up of the liquid in which the mineral has crystallized (at which point the crystallization fluid was homogeneous). The temperature at which, in the laboratory, these inclusions become homogeneous, is just the crystallization temperature (between 140 and 150°C in the fluorites studied by Calas); the permanence of the color in the mineral indicates that the crystal cannot have spent more than a time t_{max} at this temperature, and then t_{max} simply yields the formation time of the vein. The measured time for the fluorites is between 500 and 2000 y, geologically very short. This is the first time that such a determination has been accomplished.

Similar experiments using ESR and thermoluminescence in calcite have been performed recently by Ikeya and Miki [47] to date cave deposits. By ESR they were able to identify the defect formed by irradiation and responsible for the thermoluminescence; the defect which is a CO_3^{3-} radical associated with charge compensating trivalent cation impurities. Knowing the annual natural radiation dose, it is then possible to find the age and the growth velocity of the sample.

5.4. The use of point defects for optical memories [40]

5.4.1.1. Print by irradiation. Among the several methods used to store an image given by a cathode ray tube, there exists one which has already been in use for thirty years: the front part of the screen is covered with KCl (or some other alkali-halide), which is colored by the electron beam, thus fixing the image. This image may be erased by heating the support to 300°C using a resistance embedded in the screen; at this temperature, the color centers bleach out by recombining with each other. This method is very slow and is not used any more. Bishop et al. [48] have proposed high capacity two-dimensional memories which use this method. They use the luminescence properties of Tl-doped alkali-halides (see § 4.5.2). It is well known that during irradiation the luminescence associated with the Tl decreases with increasing radiation dose due to interaction of the Tl centers with the color centers thus formed. Let us then suppose that a KCl:Tl crystal be located in a camera tube against the window of a photomultiplier. Let us imagine subdividing the crystal surface in four squares corresponding to four memories. By irradiating the crystal with a very weak electron beam I_1 ($\sim 10^{-9}$ A cm^{-2} and 20 kV), the photomultiplier will yield a signal associated with the Tl luminescence on all four memories. To print an information bit (for instance on memories 1 and 3) it is sufficient to pre-irradiate the 1 and 3 zones with a moderate intensity beam I_2 ($\sim 10^{-6}$ A cm^{-2} at 20 kV); the color centers thus created will quench the Tl luminescence and thus yield a signal when the read beam will explore the crystal. To cancel, the color centers may be bleached thermally by irradiating with a high intensity beam I_3 ($\sim 10^{-4}$ A cm^{-2}: at 20 kV); the color centers will be totally destroyed and the original Tl luminescence will reappear. Bishop has given the operating times for the cycle: print ≤ 1 μs; read ≤ 100 ns; cancel ≤ 3 μs. We must remark that these times are very long, and thus this system cannot be used for fast computer memories. However its storage capacity is extremely large: 10^8 bits cm^{-2} by assuming 1 bit μ^{-2}. This last requirement implies the solution of the delicate addressing problem for the electron beam; for instance, the inhomogeneous terrestrial magnetic field may have non-negligible effects. It seems however that these difficulties may be partially overcome. Such memories could provide large capacity for library computers with small cost and fast addressing time (a magnetic tape instead has a storage capacity inferior to 10^2 bits cm^{-2}). Finally, the system discussed by Bishop et al. shows no fatigue effects after 10^5 cycles. This system also has the advantage of functioning at room temperature; this is, for instance, not the case for the methods the principle of which we shall discuss now.

5.4.1.2. Photochromic systems. Kiss [49] has reviewed several systems which may store information using light. One of the best known photochromes is rare earth (Eu^{2+} + Sm^{3+}) doped CaF$_2$. Let us call A the state CaF$_2$ + Eu^{3+} + Sm^{2+}. Divalent Sm^{2+} has two absorption bands, one in the red corresponding to electronic transitions between two localized states, the other in the blue to a transition to a conduction band state. Eu^{3+} has no absorption bands in the visible. The crystal, initially in state A and then irradiated with blue light will proceed to state B in which the electron excited away from Sm^{2+} will be trapped by Eu^{3+}: $CaF_2 + Eu^{2+} + Sm^{3+}$. Now Sm^{3+} has no absorption band in the visible, whereas Eu^{2+} has one band in the green corresponding to a transition into the conduction band. Thus irradiation with green light will bring the crystal from B back to A. The operations of this system are thus; read, using red light which can discriminate between A and B; print, with blue light which transforms A into B; erase with green light with brings B back to A. This ingenious system has nevertheless been abandoned since its efficiency is small; this implies the use of high intensity light beams and long exposure times (of the order of seconds). In fact the different transitions used correspond to intra-configurational ones ($4f \rightarrow 4f$) i.e. the relative oscillator strengths are very weak. Thus impurity concentrations must be high; however the concentration may not exceed the value at which impurity mutual interactions become important and photochromic effects disappear [50]. In other photochromic systems in alkali-halides use is made of charge or ion transfer. For instance in KCl under 200 K we have:

$$F + h\nu_F \rightarrow F^+ + F^-,$$

$$F^- + h\nu_{F^-} \rightarrow F,$$

these reactions can be used since the F^-- and F-absorption bands are separate (cf. chapter 1). This system is much more efficient than the preceding one; however it operates at low temperature.

Instead of using ionic or charge transfer, it is possible to use the reorientation properties of defects under the action of polarized light. The two systems we shall discuss presently use some particular properties of the F_A-centers [51]. This defect reorients easily; let us consider a crystal containing F_A-centers oriented randomly and illuminated with light polarized along [010] and propagating along [100]. The F_A-centers will reorient so that their symmetry axis will be perpendicular to [010]. The centers may be totally oriented and the crystal will absorb strongly in the appropriate band, light polarized along [001], whereas if the polarization is along [010] there will be no absorption (fig. 5.6). If now the crystal is illuminated with [010] light propagating in the [100] direction and with wavelength corresponding to the second transition of the F_A-center, the centers will reorient parallel to the [010]: in the second absorption band, absorption will be zero for [010] polarized light and maximum for [[001] light. Above 140 K, the reorientation efficiency is very high ($\eta \sim 2/3$), which implies a high recording sensitivity; at 55 K this efficiency practically vanishes ($\sim 3 \times 10^{-5}$); thus it is possible to "read" with polarized light; finally, in the dark, there is no thermal reorienta-

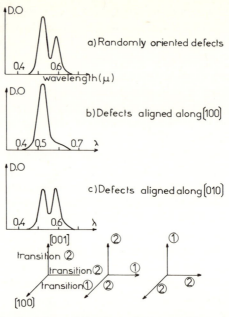

Fig. 5.6. The several orientations of F_A-center dipolar transitions and absorption spectra with light polarized along [010] and propagating along [100].

tion of the centers and thus the information may be stored indefinitely. By using the first transition Lanzl et al. [52] recorded holograms. A He–Ne laser beam is doubled in order to yield both the polarized [100] object and reference beams. The interference maxima produce a defect reorientation maximum and vice-versa. In order to "read" the holograms, the crystal is cooled to liquid nitrogen temperature and use is made of the previous light beams after removal of the object. Lanzl et al. have been able to record three-dimensional holograms of good contrast with exposure times of the order of 15 s and light power density of 1 mW cm^{-2} (corresponding to a cheap commercial laser).

Using transition (2) Blume et al. [53] have recorded holograms with better contrast. In fig. 5.7, are shown the three possible F_A-center orientations. If the crystal is illuminated with [110] polarized light in the second absorption band, nothing happens. If now a second [1$\bar{1}$0] polarized beam is used, the F_A-centers will reorient in the [010] direction. To print the holograms, the crystal is illuminated with a reference beam polarized in the [1$\bar{1}$0] direction and the object beam polarized along [110] the two beams together will induce dichroism, since the rotation of the F_A-centers will couple the two waves which would otherwise not interfere due to their perpendicular polarizations. The hologram can be read by using two beams identical to the previous ones, after cooling the crystal to liquid nitrogen temperature or below; in order to increase the signal-to-noise ratio, Blume et al. have added a polarizer whose axis was parallel to the polarization direction of the reconstruction beam.

Finally, to erase, it is sufficient to illuminate with only one beam at the printing temperature. In KCl:Na and using an argon laser (5145 Å), these authors have achieved a storage capacity of 5×10^{11} bits cm^{-3} and were able to record fifty holograms in the same crystal with an exposure time of a second. This system features a certain number of advantages: great capacity, memory stability if the temperature is kept below 0°C. However there are also certain handicaps. It is necessary to have low temperatures for the print or read parts of the holographic recording; for this however it may be hoped that a system working at room temperature be found eventually. There is another and more annoying problem: although the efficiency of the system is high ($\sim 2/3$), there is no amplification effect, i.e. a print signal corresponding to one phonon will perturb one phonon in the read part of the cycle. Now we know that it is just the amplification which makes photography based on silver salts so successful. In the photographic process the amplification factor due to the nucleation of silver colloids is 10^9. In order to compete with ordinary photography it would be necessary to find systems with this type of amplification.

Some progress has been achieved in this direction in recent years. For example Schneider et al. [54] have used the real part of the refractive index instead of the imaginary part by using F_2-centers in NaF; putting the crystal between crossed polarizers, no light may pass if all centers are randomly oriented. If in some part of the crystal the F_2-centers have a preferred orientation, the refractive index will be locally anisotropic and light will be transmitted due to depolarization. In this experiment the print mode was achieved by illuminating with near UV light, corresponding to the M_F transition; this yields the center reorientation. Reading is then done in the visible. Such a system is much more sensitive than the previous one and can work at room temperature. Much work in underway on this subject and we are quite sure that this section will be obsolete by the time this book will be in print [55].

5.5. Applications connected with defects with a permanent dipole moment

5.5.1. Low temperature dielectric thermometers

In chapter 2 we discussed the effects of para-electric impurities on the dielectric constant of a crystal and its variations with temperature or impur-

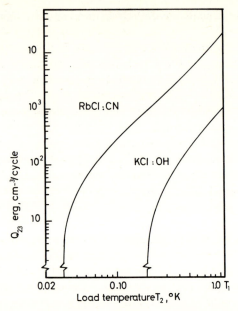

Fig. 5.7. The maximum heat quantity $Q_{2,3}$ which it is possible to extract from the cold source at temperature T_2 and to give to the hot source at temperature T_1. ($T_1 = 1.11$ K in RbCl: CN$^-$ and KCl:OH$^-$) (after Lawless [58]).

ity concentration (fig. 2.31). This characteristic makes possible low temperature thermometers with the important advantage of being insensitive to magnetic fields, contrary to the commonly used carbon or germanium thermometers. Brand et al. [56] have studied the system KCl:Li; Fiory [57] has studied KCl:OH$^-$ and KCl:CN$^-$. For sufficiently dilute systems, of a concentration high enough however to show notable para-electric effects, the crystal capacitance varies with temperature following the equation:

$$C_m = A + B/T. \tag{5.3}$$

In a KCl:Li crystal containing 3×10^{18} impurities cm^{-3}, Brand et al. have determined $A = 30$ and $B = 17.74$, with C expressed in pF and T in K. The KCl:Li works between 1 and 30 K with a sensitivity of 0.3 mK at 1 K, of 3.6 mK at 4.2 K and of 110 mK at 30 K. This thermometer is not used below 1 K since cooperative effects complicate the behavior of the dielectric constant which ceases to vary according to eq. (5.3). Nevertheless it is possible to go below 1 K by using less concentrated systems and calibrating them with known thermometers. These para-electric thermometers are not easy to use: they occupy a large volume and have large thermal inertia (for instance a KCl:OH thermometer shaped as a parallelepiped of dimensions $1 \times 1 \times 0.5$ cm^{-3} has a capacitance of 70 pF at 300 K and about 92 pF at 0.1 K); furthermore their properties change with time, since the impurities have a tendency to cluster; finally, all these crystals are somewhat hygroscopic and thus it is very difficult to avoid stray capacitances at the electrodes. However since these thermometers are of fairly recent use in fundamental research we may hope that better systems working on this principle will be found.

5.5.2. Adiabatic cooling

In particular we have shown that adiabatic cooling is possible with para-electric impurities. It would be interesting to verify if this effect would be used to build refrigerators to go below 1.2 K which would be less expensive than He^3 dilution refrigerators or adiabatic demagnetization. At this time there are no experimental devices to give an unambiguous answer to this question; thus we shall discuss instead some theoretical results of Lawless [58]. In fig. (5.7) we have shown the maximum heat extracted in a cycle from the cold reservoir and given to the hot reservoir at 1.11 K for the cases of RbCl:CN and KCl:OH, for optimum concentrations in each case. These calculations have used Carnot cycles based on the hypothesis that the applied electric field saturates the dipoles. There are then two limitations for refrigeration: first, the temperature minimum is imposed by the zero-field splitting Δ of the groundstate (see § 2.7.2) which yields $T_m = 0.07$ K for RbCl:CN and $T_m = 0.3$ K for KCl:OH; secondly, the heat transfer is limited by the dipole–dipole interaction. As stated previously, this interaction is much weaker for RbCl:CN than for KCl:OH, which explains the greater heat exchange capability of the first system. In his article, Lawless then shows that these systems are not as good as dilution cryostats, but are competitive with paramagnetic cooling. Thus these techniques may be developed to produce competitive cooling systems, and a strong applied research effort would be needed.

5.5.3. The Kerr effect

The Kerr effect consists in the difference of the indexes of refraction for light polarized parallel or perpendicular to an applied electric field in a liquid or a gas. The effect is due to the alignment of dipolar molecules along the electric field. Thus such an effect may be observed in crystals containing defects with a permanent dipole moment and which can reorient under the action of an external field. Zibold and Lüty have shown this Kerr effect in several alkali-halides doped with hydroxyl ions [59]. Such systems may yield two types of applications:

it is possible to use them as optical retarders (retardation larger than 2λ in the UV) in ultrafast UV optical modulators.

the Kerr effect extends the spectral farther than the absorption. Thus, it should be possible to use it to study para-electric defects exhibiting absorption at energies higher than the first exciton peak.

5.6. Broadly tunable lasers using color centers [60]

In 1966, Fritz and Menke [61] showed that it was possible to observe a laser emission in a flash-lamp pumped rod containing F_A (II) centers (i.e. as we said in § 1.4.1.1) F-centers associated with a small radius substitutional cationic impurity and characterized by an emission at much larger wavelength than classical F-centers (the emission of the F-center associated with Li^+ in KCl is at 0.46 eV instead of 1.12 eV for the F-center associated with Na^+ in the same salt). Until recently however the potential of color centers for useful laser action had been ignored largely due to a lack of communications between scientists from different fields. In the following we have largely used the material of the paper of Mollenauer and Olson [60].

Let us first consider the working principle of a dye laser which uses the optical pumping of F-centers (fig. 5.8), yielding a four-level system. The optical pumping consists of four steps; excitation, relaxation of excited state, emission and relaxation to the groundstate. τ_R and τ_R' are very short with respect to the luminescence decay time and the only populations of any significance are N and N^*, that is the populations of the relaxed groundstate

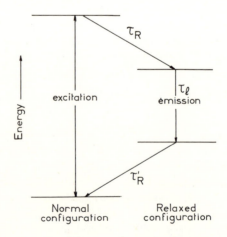

Fig. 5.8. The optical pumping cycle of F-centers.

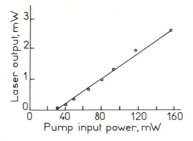

Fig. 5.9. Output versus input power for a KCl:Li laser; output mirror transmission 1.6%, total cavity loss 7.4% (after Mollenauer and Olson [60]).

and excited states respectively. If there is no absorption of the emitted light by the crystal, the gain coefficient α_0 at the band peak for a Gaussian band of full width at the half-maximum δv is given by the formula:

$$\alpha_o = (N^* \lambda_0^2 \eta / 8\pi n^2 \tau_1) \, (1/1.07 \, \delta v), \tag{5.4}$$

where λ_0 is the wavelength at the band center, n is the host index, η is the quantum efficiency and τ_1 is the measured luminescence decay time. Table 5.2, gives the gain calculated by Mollenauer and Olson for three color centers having the 4-level scheme (like practically all color centers), F-centers, F_A (II) (KCl:Li) centers and F_2^+ centers which are two adjacent halogen vacancies along $\langle 110 \rangle$ with one electron.

From this table, it is clear that large gain can be reached with F_2^+- and F_A(II)-centers. The stimulated emission of F-centers in KCl has been observed by De Martini et al. [62] but the last two last systems are more promising for the following three reasons:

There is practically no absorption in the energy range of their emission whereas the F-center emission can be reabsorbed by F^--centers formed during the pumping. The two other defects are very stable under the action of the pumping light and the expected gain is much larger.

Table 5.2
Values of α_0 for the F- and F_2^+-centers in KCl and for the F_A (II) center in KCl:Li. $N^* = 10^{16} \, cm^{-3}$ in all cases (from Mollenauer and Olson [60].

Quantity	F	F_2^+	F_A(II)	Units
λ_0	1	1.68	2.7	μm
τ_1/η	600	200	200	ns
δv	6.3	1.69	1.45	10^{13} Hz
α_0	0.004	35	4.2	cm^{-1}

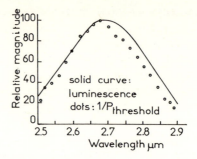

Fig. 5.10. Tuning characteristics of the KCl:Li laser (after Mollenauer and Olson [60]).

Figures 5.9, 5.10 and 5.11 show the results obtained, the crystal being pumping by a krypton ion laser operating at 6471 Å. In this experiment the crystal slab was held against a Cu cold finger (77 K) with a gentle string clamp with no grease or other thermally conductive compounds in order to avoid any strain or fracture of the crystal. Despite the rather poor thermal contact between the crystal and the cold finger, CW pump inputs as high as about 200 mW could be tolerated without undue heating of the crystal.

As Mollenauer and Olson have stated, this new field is very promising for the following reasons:

color centers are covering a very large range of frequencies. (In their work, they have not explored a lot of other systems like F_2^+ in LiF and NaF, F_3^+ in different salts etc . . .)[†]

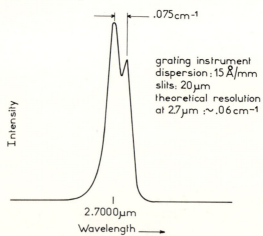

Fig. 5.11. Spectral purity of the KCl:Li laser output (after Mollenauer and Olson [60]).

† Other systems have been more recently explored [64].

they can provide tunable lasers in an energy range that is not well covered and they would provide inexpensive and practical sources for molecular spectroscopy, pollution detection etc . . .

this laser has a static amplifying medium as opposed to a rapidly flowing and turbulent dye solution.

In conventional dye laser experiments as well as in the experiment described here, a dispersive system (or a filter) is fixed in the cavity to allow light amplification only at a given wavelength. Bjorklund et al. [63] have recently shown that it is possibly to make a very inexpensive laser emitting at a predetermined frequency using the principle of distributed feedback which is the following: if either the gain coefficient α_0 or the refractive index n, is modulated spatially with period d, there will be strong feedback at those wavelengths that satisfy the Bragg condition:

$$n\lambda = 2d$$

without the need of external mirrors. In this last experiment they have modulated spatially the F_A (II) concentration using the photochromic conversion process in a KCl crystal doped with lithium and hydrogen:

$$U \to F \to F_A(II).$$

The $U \to F$ conversion occurs by irradiation in the U-band in the UV range and the modulation was performed by exposing the crystal samples at room

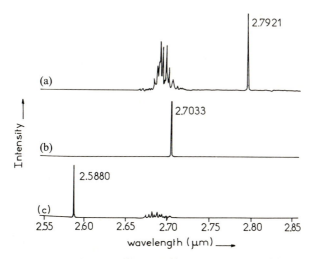

Fig. 5.12. Output spectra of three different KCl:Li laser samples with a modulated concentration of F_A (II) centers. Sample (a), fringe spacing $0.9483\,\mu$m. Sample (b), fringe spacing $0.9185\,\mu$m. Sample (c), fringe spacing $0.8804\,\mu$m (after Bjorklund et al. [63]).

temperature to interfering beams of intense 266 nm radiation (the fourth harmonic of a Nd:YAG laser). In this experiment, the cavity is the crystal itself which is placed in a liquid nitrogen immersion Dewar and pumped longitudinally or transversally at 532 nm with the second harmonic of the Nd:YAG laser. The spectral output that they obtained is shown on fig. 5.12 with an efficiency of 6.7%.

It was quite clear to us in reading the papers of Mollenauer and his co-workers that this application of color centers would have a great future.

References to chapter 5

[1] J. H. Crawford. Jr. Adv. Phys. **17** (1968) 93.
[2] E. Sonder and W. A. Sibley, in: Point Defects in Solids, ed. J. H. Crawford. Jr. and L. M. Slifkin (Plenum Press, New York, 1972).
[3] Y. Farge, J. de Phys. Suppl. **C34** (1973) 475.
[4] M. N. Kabler, in: Radiation Damage Processes in Materials, Ed. C. H. S. Dupuy, (Noordhoff International, Leyden, 1975).
[5] N. Itoh, Nucl. Inst. and Methods **132** (1976) 201; M. Saidoh and P. D. Townsend, Radiat. eff. **27** (1975) 1.
[6] T. P. Hall, A. E. Hughes and D. Pooley, J. Phys. C: Solid St. Phys. **9** (1976) 439.
[7] P. Durand, Y. Farge and M. Lambert, J. Phys. Chem. Solids **30** (1969) 1352.
[8] E. Sonder, Phys. Rev. **B12** (1975) 1516.
[9] G. Guilhot, A. Nouilhat, and P. Pinard, J. Phys. Soc. Japan **39** (1975) 398.
[10] N. Ioth and N. Saidoh, J. Phys. C: Solid St. Phys. **34** (1973) 101; M. Saidoh and N. Itoh, J. Phys. Chem. Solids **34** (1973) 1165.
[11] Y. Kondo, M. Hirai and M. Ueta, J. Phys. Soc. Japan **33** (1972) 151 and refs. therein.
[12] A. Schmid, P. Bräunlich and P. K. Rol, Phys. Rev. Letters **35** (1975) 1382.
[13] R. Smoluchowski, Phys. Rev. Letters **35** (1975) 1385.
[14] P. D. Townsend, R. Browning, D. J. Garlant, J. C. Kelly, A. Mamjoubi, A. J. Michael and M. Saidoh, Radiat. Eff. **30** (1976) 55.
[15] J. N. Bradford, R. T. Williams and W. L. Faust, Phys. Rev. Letters **35** (1975) 300.
[16] R. T. Williams, Phys. Rev. Letters **36** (1976) 529.
[17] Y. Toyozawa, Int. Conf. Color Centers (Sendai, 1974) J. Phys. Soc. Japan **44** (1978) 482.
[18] W. B. Fowler, in: Physics of Colors Centers (op. cit) P. 123.
[19] J. H. Crawford. Jr, Proc. Conf. Color Center Crystal Luminescence (Turin, 1960).
[20] F. Lüty, in: Int. Conf. Color Centers, (Reading, 1971) abstract H158.
[21] T. Karazawa and M. Hirai, J. Phys. Soc. Japan **40** (1976) 769.
[22] M. Ikeda, N. Itoh, T. Okada and T. Suita, J. Phys. Soc. Japan **21** (1966) 1304; Y. Farge, J. Phys. Chem. Solids **30** (1966) 1375; See also: Y. Susuki and M. Hirai, J. Phys. Soc. Japan **43** (1977) 1679; J. M. Ortega, to be published.
[23] E. Sonder, Phys. Rev. **B5** (1972) 3259.
[24] Y. Farge, M. Lambert and R. Smoluchowski, Solid St. Commun. **4** (1966) 333; Phys. Rev. **159** (1967) 700.
[25] J. L. Paul and A. B. Scott, Phys. Stat. Solidi (b) **52** (1972) 581.
[26] I. Schneider, Sol. St. Commun. **9** (1971) 2191.
[27] D. Pooley, Proc. Phys. Soc. **89** (1966) 723.
[28] D. K. Dawson and D. Pooley, Sol. St. Commun. **7** (1969) 1001.
[29] D. Pooley, in: Radiation Damage processes in Materials. Ed C. H. S. Dupuy (Noordhoff, Leyden, 1975).

[30] Y. Farge, Phys. Rev. **B1** (1970) 4797.
[31] Selected Topics in Radiation Dosimetry (Int. Atomic Energy Agency, Vienna 1961); Proc: Int. Conf. on luminescence Dosimetry (Stanford, 1965).
[32] A. E. Hughes and D. Pooley, Real Solids and Radiation, (Wykeham Publications London, 1975) p. 168.
[33] J. H. Schulman, F. H. Attix, E. J. West and R. J. Ginther, see ref. [31] p. 531.
[34] C. C. Klick, E. W. Claffy, S. G. Gorbies, F. H. Attix, J. H. Schulman and J. C. Allard, J. Appl. Phys. **38** (1967) 3867.
[35] M. R. Mayhugh, R. W. Christy and N. M. Johnson, J. Appl. Phys. **41** (1970); 2968; M. R. Mayhugh, J. Appl. Phys. **41** (1970) 4776; J. H. Jackson and A. M. Hans, J. of Phys. C: 3 Solid State Phys. (1970) 1967.
[36] M. C. Wintersgill, P. D. Townsend and F. Cusso-Perez, J. Phys. C: Solid St. Phys. **7** (1976) 123.
[37] R. Nink and H. J. Kos, J. Phys. **37** C: Solid St. Phys. **7** (1976) 127.
[38] V. H. Ritz, A. E. Nash and F. H. Attix, in: 4th Int. Symp. on Exoelectron Emission and Dosimetry (Liblice, Czechoslovakia, 1973) Czah. Acad. Sci. (1974) 258.
[39] E. C. McCullough, G. D. Fullorton and J. R. Cameron, J. Appl. Phys. **43** (1972) 77.
[40] D. Pooley, in: Radiation Damage Processes in Materials, ed. C. H. S. Dupuy (Noordhoff Int., Leyden, 1975).
[41] G. Calas, Thesis 3th cycle, Université Paris VI (1971) unpublished.
[42] K. Przibram, Verfarlüng und lumineszenz, Beitrage son Mineralphysiln (Springer Verlag, 1953).
[43] H. Bill, J. Sierro and R. Lacroix, Am. Min. **52** (1967) 1003.
[44] G. A. Il'inski, Zav. Uses. Min. **91** (1962) 613.
[45] S. R. Titley and P. E. Damon, J. Geophys. Res. **67** (1962) 4491.
[46] G. Calas, H. Curien, Y. Farge and G. Maury, C. R. Acad, Sci. **274** (1972) 781; G. Calas, Bull. Soc. Frans. Min. Crystal **95** (1972) 470.
[47] M. Ikeya and T. Miki, Int. Conf. Defects in Insulating Crystals (Gatlinsburg, 1977).
[48] H. E. Bishop, R. P. Hendorson, A. E. Hughes, P. Iredale and D. Pooley, Int. Con. Color Centers (Sendai, 1974 abstract no. 188.
[49] Z. J. Kiss, Phys. Today **23** (1972) 42.
[50] E. Spitz, private communication.
[51] F. Lüty, in: Physics of Color Centers (op. cit.) ch. 3.
[52] F. Lanzl, U. Roder and W. Waidelich, Appl. Phys. Letters **18** (1971) 56.
[53] H. Blume, T. Bader and F. Lüty, Int. Conf on Color Centers (Reading, 1971) abstract 136.
[54] I. Schneider, M. Lehmann, and R. Bocker, Appl. Phys. Letters **25** (1974) 77.
[55] D. Casasent and F. Caimi, Appl. Phys. Letters. **29** (1976) 660; Appl. Optics **15** (1976) 815; I. Schneider and M. E. Gingerich, Appl. Optics **15** (1976) 2429.
[56] R. A. Brand, S. A. Letzring, H. S. Sack and W. W. Webb, Rev. Sci. Int. **42** (1971) 927.
[57] A. T. Fiory, Rev. Sci. Int. **42** (1971) 930.
[58] W. N. Lawless, J. Appl. Phys. **40** (1969) 4448.
[59] G. Zibold and F. Lüty, Int. Conf. Color Centers (Reading, 1971) abstract 71; F. Lüty and A. Diaz-Gongora, Phys. Stat. Solidi **85** (1978) 693; A. Diaz-Gongora and F. Lüty, Phys. Stat. Solidi **86** (1978) 482.
[60] L. F. Mollenauer and D. H. Olson, J. Appl. Phys. **46** (1979) 3109.
[61] B. Fritz and E. Menke, Solid St. Commun. 3 (1965) 61.
[62] F. de. Martini, U. M. Grassano and F. Simoni, Int. Conf. Color Centers (Sendai, 1974) abstract A7.
[63] G. C. Bjorklund, L. F. Mollenauer and W. J. Tomlinson, Appl. Phys. Letters **29** (1976) 116.

AUTHOR INDEX

SUBJECT INDEX

PREVIOUSLY PUBLISHED BOOKS IN THIS SERIES: